DEFENSE, SECURITY AND STRATEGIES

COMBATANT COMMANDS IN THE UNITED STATES ARMED FORCES

ELEMENTS AND CONSIDERATIONS

DEFENSE, SECURITY AND STRATEGIES

Additional books in this series can be found on Nova's website under the Series tab.

Additional E-books in this series can be found on Nova's website under the E-book tab.

AMERICAN POLITICAL, ECONOMIC, AND SECURITY ISSUES

Additional books in this series can be found on Nova's website under the Series tab.

Additional E-books in this series can be found on Nova's website under the E-book tab.

DEFENSE, SECURITY AND STRATEGIES

COMBATANT COMMANDS IN THE UNITED STATES ARMED FORCES

ELEMENTS AND CONSIDERATIONS

QUINCY N. BUTLER
AND
HARRIS A. SIMMONS
EDITORS

Nova Science Publishers, Inc.
New York

Library of Congress Cataloging-in-Publication Data

ISBN: 978-1-61942-924-6

Published by Nova Science Publishers, Inc. † New York

Contents

PREFACE

The Unified Command Plan (UCP) and associated Combatant Commands (COCOMs) provide operational instructions and command and control to the Armed Forces and have a significant impact on how they are organized, trained and resourced - areas over which Congress has constitutional authority. The UCP is a classified executive branch document prepared by the Chairman of the Joint Chiefs of Staff and reviewed and updated every two years that assigns missions; planning, training, and operational responsibilities; and geographic areas of responsibilities to COCOMs. This book provides information on the history, mission, and operational considerations of the UCP and COCOMs with a focus on current issues and considerations.

Chapter 1 - The Unified Command Plan (UCP) and associated Combatant Commands (COCOMs) provide operational instructions and command and control to the Armed Forces and have a significant impact on how they are organized, trained, and resourced—areas over which Congress has constitutional authority. The UCP is a classified executive branch document prepared by the Chairman of the Joint Chiefs of Staff (CJCS) and reviewed and updated every two years that assigns missions; planning, training, and operational responsibilities; and geographic areas of responsibilities to COCOMs. Functional COCOMs operate world-wide across geographic boundaries and provide unique capabilities to geographic combatant commands and the Services while Geographic COCOMs operate in clearly delineated areas of operation and have a distinctive regional military focus.

Chapter 2 - It is my privilege as Commander of United States Africa Command to present to Congress our Posture Statement for 2011. United States Africa Command protects American lives and promotes our national interests by advancing security and stability in Africa. We follow two main lines of effort building the security capacity of our African partners and preparing for a wide range of potential crises. U.S. Africa Command's operations, programs, and activities contribute to reducing the threats to our citizens and interests both abroad and at home by helping African states provide for their own security. We seek to enhance regional stability through support to and partnership with African regional organizations. Our planning and training are designed to prevent conflict while simultaneously ensuring that U.S. Africa Command is prepared to respond decisively to any crisis when the President so directs.

Chapter 3 - U.S. Central Command (CENTCOM) oversees operations alongside our allies, partners, and friends in a critically important region of the world. CENTCOM is engaged throughout the greater Middle East and South Central Asia across the full spectrum of warfare, standing against violent aggression and the tyranny of militant extremists, while contributing to the broader conditions for peace, stability, and prosperity.

Chapter 4 - Mr. Chairman, Ranking Member, and distinguished members of the committee, I would like to thank you for the opportunity to appear before you today to share with you the successes achieved and the challenges being faced by the men and women of both the United States European Command and the North Atlantic Treaty Organization's (NATO) Allied Command Operations since I last appeared before you. I have now been at the helm of these commands for almost two years and am happy to report we continue to make progress and develop stronger partnerships for our shared security. The most important activities and initiatives contained in these pages are those in which we work together with our allies and partners to build capacity to ensure U.S. security in the European theater and, thus, defend our homeland forward.

Chapter 5 - Chairman Levin, Senator McCain, distinguished members of the committee, thank you for the opportunity to share my thoughts on the missions and focus areas of U.S. Northern Command (USNORTHCOM) and North American Aerospace Defense Command (NORAD). Let me begin by saying how impressed I am by the talented cadre of professionals—from the active and reserve components of all Services of the U.S. military, DOD civilians, our Canadian partners, and representatives from 68 different civilian organizations—who work in and about my headquarters in Colorado Springs each and every day. It is truly humbling to work with such a great team.

Chapter 6 - Chairman Levin, Senator McCain and distinguished members of the committee, thank you for this opportunity to provide an update on United States Pacific Command (USPACOM). Today is my second testimony as the Commander of USPACOM. Since the 2010 hearings, much has happened in this increasingly important region. I have travelled throughout the Asia-Pacific visiting a dozen different countries—some multiple times—and gained valuable insight into the extensive contributions by our regional allies and partners.

Chapter 7 - Chairman Levin, Senator McCain, and distinguished members of the committee: thank you for the opportunity to appear before you today to discuss United States Southern Command's accomplishments and future efforts in Central and South America and the Caribbean. Over the past year, we worked in close collaboration with U.S. government agencies and our partner nations to respond to the unprecedented natural disaster in Haiti and to the ongoing threats to regional security. This year, with the support of Congress, we will build on our accomplishments and continue to foster close cooperation and engagement throughout the region. We will also continue to evolve as a joint and interagency organization that promotes U.S. national and regional security interests through enduring partnerships.

Chapter 8 - Mr. Chairman and distinguished members of the committee, thank you for this opportunity to provide an update on the United States Special Operations Command (USSOCOM). Our Special Operations Forces give us much cause for great pride and it is my deep privilege to represent them to you, and especially to do so for the fourth time as their commander.

Chapter 9 - Chairman Levin, Senator McCain, and members of the committee, thank you for the opportunity to present my views on United States Strategic Command's (USSTRATCOM) missions and priorities. Today marks my first appearance before you as the Commander of USSTRATCOM. I have occupied this position for a short time, but I have been involved with the Command's missions for much of my career. I was privileged to spend two years as the Deputy Commander, and most recently I served as one of USSTRATCOM's Service component commanders. USSTRATCOM's active duty, reserve, and civilian members, who are standing watch this very minute at locations across the country and around the globe, exemplify the best of today's joint force. I am privileged to lead this remarkable team, and I look forward to working with you to assure our nation's security.

Chapter 10 - The United States Transportation Command (USTRANSCOM) is a Total Force team of Active Duty, Guard, Reserve, civilian, commercial partners and contractors who lead a world class Joint Deployment and Distribution Enterprise (JDDE). Through our component commands, the Army's Military Surface Deployment and Distribution Command (SDDC), the Navy's Military Sealift Command (MSC), the Air Force's Air Mobility Command (AMC) and our commercial partners, we execute military and commercial transportation, distribution process integration, terminal management, aerial refueling and global patient movement across the full range of military and humanitarian operations. We operate effectively and efficiently to deploy and sustain the warfighter...and 2010 was a banner year. Together with our Service, Combatant Command, Interagency and Coalition partners, the USTRANSCOM team responded superbly to the President's direction to increase forces by 30,000 in Afghanistan, to drawdown forces to 50,000 in Iraq and to an unprecedented series of world events and natural disasters. Whether delivering combat power to Afghanistan through logistics or humanitarian relief to the people of Pakistan, Haiti and Japan, our team kept our promises and delivered on time, on target and at best value to the taxpayer.

In: Combatant Commands in the United States Armed Forces ISBN: 978-1-61942-924-6
Editors: Quincy N. Butler and Harris A. Simmons

Chapter 1

THE UNIFIED COMMAND PLAN AND COMBATANT COMMANDS: BACKGROUND AND ISSUES FOR CONGRESS*

Andrew Feickert

SUMMARY

The Unified Command Plan (UCP) and associated Combatant Commands (COCOMs) provide operational instructions and command and control to the Armed Forces and have a significant impact on how they are organized, trained, and resourced—areas over which Congress has constitutional authority. The UCP is a classified executive branch document prepared by the Chairman of the Joint Chiefs of Staff (CJCS) and reviewed and updated every two years that assigns missions; planning, training, and operational responsibilities; and geographic areas of responsibilities to COCOMs. Functional COCOMs operate world-wide across geographic boundaries and provide unique capabilities to geographic combatant commands and the Services while Geographic COCOMs operate in clearly delineated areas of operation and have a distinctive regional military focus. There are currently nine COCOMs:

- USSOCOM: U.S. Special Operations Command, MacDill Air Force Base, FL.
- USSTRATCOM: U.S. Strategic Command, Offutt Air Force Base, NE.
- USTRANSCOM: U.S. Transportation Command, Scott Air Force Base, IL.
- USAFRICOM: U.S. Africa Command, Kelley Barracks, Stuttgart, Germany.
- USCENTCOM: U.S. Central Command, MacDill Air Force Base, FL.
- USEUCOM: U.S. European Command, Patch Barracks, Stuttgart, Germany.
- USNORTHCOM: U.S. Northern Command, Peterson Air Force Base, CO.
- USPACOM: U.S. Pacific Command, Camp H.M. Smith, HI.
- USSOUTHCOM: U.S. Southern Command, Miami, FL.

* This is an edited, reformatted and augmented version of a Congressional Research Service publication, CRS Report for Congress R42077, prepared for Members and Committees of Congress ,dated November 7, 2011.

This report provides information on the history, mission, and operational considerations for each of these organizations as well as a brief discussion of current issues associated with the UCP and these commands.

The origins of the UCP and COCOMs are rooted in World War II. After the war, U.S. leaders, taking advantage of the lessons learned in both theaters, initiated a series of legislative changes that resulted in the current UCP process and COCOM construct.

The UCP and COCOMs are covered under Title 10 - Armed Forces; Subtitle A - General Military Law; Part I–Organization and General Military Powers; Chapter 6–Combatant Commands. These provisions detail the responsibilities and authorities of COCOMs as well as legal requirements related to the UCP.

A potential issue for Congress is whether there is a need for greater interagency involvement in the UCP development process. Another possible area for congressional concern is if Geographical COCOMs have made U.S. foreign policy "too militarized." Some have also suggested there might be a need for separate COCOMs apart from the current nine to better address emerging regional and ethnic alignments as well as emerging threats such as cyber warfare. Finally, if Congress believes the current COCOM construct does not meet contemporary or future security requirements, there are proposals for alternative organizational structures that might prove more effective.

What Are the Unified Command Plan (UCP) and Combatant Commands (COCOMs)?

Introduction

The Unified Command Plan (UCP) and associated Combatant Commands (COCOMs) provide operational instructions and command and control to the Armed Forces and have a significant impact on how they are organized, trained, and resourced—areas over which Congress has constitutional authority.

In a grand strategic sense, the UCP and the COCOMs are the embodiment of U.S. military policy both at home and abroad. The COCOMs not only execute military policy but also play an important role in foreign policy, and Congress, in both oversight and budgetary roles, has shown great concern in this regard.

All Combatant Commanders testify to the Armed Services Committees on an annual basis about their posture and budgetary requirements and frequently host Members and staff during a variety of congressional delegation visits.

Unified Command Plan (UCP)

The Department of Defense (DOD) defines the Unified Command Plan (UCP) as

> The document, approved by the President, that sets forth basic guidance to all unified combatant commanders; establishes their missions, responsibilities, and force structure; delineates the general geographical area of responsibility (AOR) for geographic combatant commanders; and specifies functional responsibilities for functional combatant commanders.[1]

The UCP is a classified executive branch document prepared by the Chairman of the Joint Chiefs of Staff (CJCS) and reviewed and updated at a minimum every two years. While the UCP is normally on a two-year cycle, it can be updated anytime based on changing strategic, political, and budgetary requirements. As noted, the UCP assigns missions; planning, training, and operational responsibilities; and geographic areas of responsibilities to COCOMs. The UCP is assessed and modified, taking into consideration the following strategic documents:[2]

- The National Security Strategy of the United States of America;
- The National Defense Strategy of the United States of America;
- The National Military Strategy of the United States of America; and
- The current UCP.

The UCP process also takes into consideration the strategic context (such as the war in Afghanistan, the global economic situation, relationships with allies, etc.) and command guidance from the President and senior DOD civilian and military leadership. As part of the final review process before the UCP is submitted to the President, the proposed UCP is reviewed by senior service leaders, the Secretary of Defense, and the National Security Council (NSC). Congress is not included in this review process but does have visibility into issues affecting UCP development.

Combatant Command (COCOM)

DOD defines Combatant Command (COCOM) as:

> A unified[3] or specified[4] command with a broad continuing mission under a single commander established and so designated by the President, through the Secretary of Defense and with the advice and assistance of the Chairman of the Joint Chiefs of Staff. Combatant commands typically have geographic or functional responsibilities.[5]

Dr. Cynthia Watson, a professor at the National War College and author of "Combatant Commands: Origins, Structure, and Engagements" describes combatant commands as being:

> Commands in charge of utilizing and integrating air, land, sea, and amphibious forces under their commands to achieve U.S. national security objectives while protecting national interests. Four of the unified commands handle functional concerns while there are six with geographic mandates. The specific configurations have shifted over the decades, but the idea that geography provides an organizing principle remains the same, allowing each combatant command to have its specific threats and opportunities. The combatant commanders work with the military forces in their theaters, and report to the commander in chief and secretary of defense. The combatant commanders do not serve on the Joint Chiefs of Staff nor are they the senior U.S. representatives in the theater.[6]

The number of combatant commands is not regulated by law or policy and their numbers and responsibilities have varied over the years.

Today, there are nine active COCOMs and one COCOM–U.S. Joint Forces Command (JFCOM)—is currently in the process of being disestablished. JFCOM is included in this report because there has been congressional concern about how its functions are being re-distributed throughout DOD.

Functional Combatant Commands

Functional combatant commands operate world-wide across geographic boundaries and provide unique capabilities to geographic combatant commands and the Services:

- USJFCOM: U.S. Joint Forces Command, Norfolk Naval Base, VA. (*disestablished August 2011*) ;
- USSOCOM: U.S. Special Operations Command, MacDill Air Force Base, FL;
- USSTRATCOM: U.S. Strategic Command, Offutt Air Force Base, NE; and
- USTRANSCOM: U.S. Transportation Command, Scott Air Force Base, IL.+

Geographic Combatant Commands

Geographic combatant commands operate in clearly delineated areas of operation and have a distinctive regional military focus.

- USAFRICOM: U.S. Africa Command, Kelley Barracks, Stuttgart, Germany;
- USCENTCOM: U.S. Central Command, MacDill Air Force Base, FL;
- USEUCOM: U.S. European Command, Patch Barracks, Stuttgart, Germany;
- USNORTHCOM: U.S. Northern Command, Peterson Air Force Base, CO;
- USPACOM: U.S. Pacific Command, Camp H.M. Smith, HI; and
- USSOUTHCOM: U.S. Southern Command, Miami, FL.

Origins of the UCP

The United States' experience with global warfare in World War II provided countless lessons attesting to the importance of unity of military effort achieved through the unified command of U.S. forces.[7]

While the United States was able to achieve a degree of unified command in the European theater under General Dwight Eisenhower—Supreme Commander, Allied Expeditionary Force—attempts to establish unified command in the Pacific "proved impossible."[8] Differences between the Army and Navy precluded any sort of unified command arrangement and General Douglas MacArthur commanded U.S. Army Forces, Pacific while Admiral Chester Nimitz commanded the U.S. Pacific Fleet. Although both commanders were able to work together to defeat Japan, there was a considerable amount of friction between these two powerful, independent commands. After the war, President Truman noted:

> We must never fight another war the way that we fought the last two. I have the feeling that if the Army and Navy had fought our enemies as hard as they fought each other, the war would have ended much earlier.[9]

Outline Command Plan 1946—The First UCP10

In 1946, the Chief of Naval Operations characterized the Pacific command arrangement as "ambiguous and unsatisfactory" and proposed a single command over the Pacific (not including Japan, Korea, and China) to provide unity of command over all U.S. forces in the region. The Army and the Army Air Forces rejected this proposal, favoring instead unified command based on assignment of mission and forces as opposed to geographic areas. After a great deal of discussion and compromise, a worldwide system of unified command was established. President Truman approved the "Outline Command Plan" in December 1946 which established seven commands as an "interim measure for the immediate post war period." The seven commands were:

- Far East Command;
- Pacific Command;
- Alaskan Command;
- Northeast Command;
- Atlantic Fleet;
- Caribbean Command; and
- European Command.

Some of these seven commands contained more than one service and were, in a sense, unified, while others, such as the Atlantic Fleet, were service-specific. Even though these commands were established to achieve a degree of unity, the Services continued in many instances to plan and act independently. Since 1946, the UCP has continued to evolve—sometimes in a dramatic manner— to reflect ever changing strategic, organizational, and political requirements. While Congress has influenced the UCP over the years, three major legislative initiatives have had a lasting impact on the UCP.

National Security Act of 1947 (P.L. 80-253)

While the National Security Act of 1947 is best known for the creation of the U.S. Air Force, the Central Intelligence Agency (CIA), and establishing the office of the Secretary of Defense, it also created the Unified Combatant Command (UCC) system.[11] The UCC system signified the recognition by the United States that it would continue to have a world-wide, continuous global military presence. The National Security Act of 1947 also gave the Joint Chiefs of Staff (JCS) the responsibility to establish unified commands in "strategic areas" subject to the approval of the President and Secretary of Defense.[12]

DOD Reorganization Act of 1958 (P.L. 85-599)[13]

In 1958, President Eisenhower—the former Supreme Commander, Allied Expeditionary Force— decided a more unified and streamlined chain of command to employ combat forces was needed, essentially putting an end to separate land, sea, and air combat. President Eisenhower sought "a complete unification of all military planning and combat forces and commands" and proposed the DOD Reorganization Act of 1958 to Congress to amend the National Security Act of 1947.

The DOD Reorganization Act of 1958 authorized the President, acting through the Secretary of Defense with the advice of the JCS, to establish unified or specified commands, assign missions, and determine their force structure. This act did not alter any of the authorities established by the National Security Act of 1947 but instead established a clear line of command from the President, through the Secretary of Defense, to the combatant commanders. Combatant commanders were delegated full operational control over forces assigned to them but once these forces were assigned, they could only be transferred with presidential approval. Responsibility for the administration of these assigned forces was to remain with their respective Services.

Goldwater-Nichols DOD Reorganization Act of 1986 (P.L. 99-433)[14]

In the aftermath of the failed 1980 multi-service mission to rescue U.S. hostages in Iran and the 1983 invasion of Grenada which featured numerous instances of poor inter-service planning and cooperation, there was renewed emphasis on "jointness" both in Congress and at the Pentagon. Goldwater-Nichols sought to "rebalance the relative power of the geographic commands versus the services."[15] Goldwater-Nichols called for the Chairman of the JCS (CJCS) to review the missions, responsibilities, and force structure and geographic boundaries for each COCOM not less than every two years and recommend changes to the Secretary of Defense and the President. In addition, the act expanded the CJCS's and combatant commander's powers and gave combatant commanders greater interaction with Congress and greater participation in the DOD budget process.

What Laws Govern the UCP?

The UCP and COCOMs are covered under Title 10 - Armed Forces; Subtitle A - General Military Law; Part I–Organization and General Military Powers; Chapter 6–Combatant Commands. As it relates to the UCP, Section 161, inter alia stipulates:

- Unified COCOMs are established by the President, through the Secretary of Defense (SECDEF), with the advice and assistance of the CJCS.
- The CJCS shall periodically review (at least every two years) missions, responsibilities (including geographic boundaries), and force structure of each combatant command.
- Based on this review, the CJCS will recommend to the President, through the SECDEF, changes to missions, responsibilities, and force structure deemed necessary.
- The President, except in times of hostilities or imminent danger, will notify Congress not less than 60 days after establishing a new combatant command or significantly revising the missions, responsibilities, or force structure of an existing combatant command.

Also under Section 161, the CJCS is required to consider during each periodic UCP review:

- Whether there was an adequate distribution of responsibilities among the regional unified combatant commands;

- Whether fewer or differently configured commands would permit the United States to better execute war fighting plans;
- Whether any assets or activities were redundant;
- Whether war fighting requirements were adequate to justify current commands;
- Whether exclusion of certain nations from the Areas of Responsibility presented difficulties with respect to national security objectives in those areas; and
- Whether the boundary between the U.S. Central and European Commands could create command conflicts in the context of a major regional conflict in the Middle East.

UCP UPDATE CYCLE[16]

Generally, the UCP update cycle runs from 12 to 18 months. The current UCP process consists of five iterative phases described below:

1. *Guidance*: DOD UCP planners review four central documents: The National Security Strategy of the United States of America; The National Defense Strategy of the United States of America; The National Military Strategy of the United States of America; and the current UCP. UCP participants also receive command guidance in various forms and in various degrees, and are apprised of the strategic context under which the UCP will be evaluated. During the final part of this phase, stakeholders (Combatant Commanders, Service Chiefs, and the Joint Staff) identify issues that they believe need to be addressed during the UCP update cycle.
2. *Slate:* During this phase, issues are slated for discussion. At the beginning of the phase an action officer planning conference is held to discuss UCP issues. After the conference, stakeholders submit Issue Development Papers (IDPs) to the Joint Staff where they are posted online on a secure operating system
3. (SIPRNET)[17] where stakeholders can view and comment on them. After a period of time, these IDPs and associated comments are sent to the Deputy Director forStrategy and Policy and the Director J-5 for validation. Those IDPs that make it through the validation process are then "slated" and placed online so that stakeholders can develop positions for the next phase of the UCP process.
4. *Assessment:* This phase begins with a Planner's Assessment Conference where courses of action (COA) are developed for each validated IDP. These COA provide decision makers with a range of choices to address the IDPs. Once the COA are agreed at the conference, they are again posted online for review and comment. After a period of time, the COA are finalized and "closed out."
5. *Adjudication*: The adjudication phase is a four part process whereby the issues and COA are sent to various levels of the Joint Staff for approval. The first level is the Deputy Director for Strategy and Policy and after review and approval, a draft UCP is published. Next comes the Director J-5 and another revised draft UCP is published. This draft UCP is then taken to a Service Chiefs' "Tank"[18] meeting and after that meeting, another UCP draft is prepared. Finally a Joint Chiefs of Staff Tank

meeting is held and a final draft UCP is prepared, posted on the SIPRNET, and is then ready for final review.

6. *Review*: During this phase the UCP is reviewed and revised for the final time. The first review is held at the "four-star level" including Service Chiefs, Combatant Commanders, and other 4-star level general officers and DOD civilians. Next, the SECDEF reviews the draft UCP and suggests changes. The next step is the National Security Council (NSC) review where the UCP is commented on by other U.S. government agencies. Finally, after incorporating the views of the NSC principals, the UCP is taken to the President for approval and final publication.

OTHER AGENCY INVOLVEMENT[19]

There are other executive branch agencies (State Department, Department of Justice, Department of Homeland Security, and the Central Intelligence Agency, to name a few) that are collectively referred to as the Interagency and have a vested interest in the UCP because some of its associated military tasks intersect with the responsibilities of these agencies. While none of these agencies are formally part of the UCP development process, they do have access to it by means of agency liaison officers stationed at the COCOMs and on the Joint Staff.

These liaison officers have visibility of the IDP and COA process as well as access to draft UCPs and are able to report their observations and concerns to their principals (i.e. Secretary of State, Attorney General, etc.). The NSC also receives periodic updates on UCP development or revisions during the UCP cycle. The principals may then choose to address their UCP concerns with senior DOD leadership. During the NSC UCP Review, other agencies can publicly voice concerns with the UCP but, unless an agency has not been actively following the UCP development, there should be no "surprises" when the UCP is reviewed by the NSC.

CONGRESSIONAL INVOLVEMENT[20]

Congress currently has no statutory role in the UCP development, revision, or review process other than those stipulated in Title 10, Chapter 6, Sections 161 and 166. Congress does, however, make its concerns known during hearings, private conversations between Members and staff and DOD leadership, and through lending support to UCP-related issues through legislation or by resolution. As a relatively recent example, prior to the 2007 decision to stand up AFRICOM, a number of Members called for the creation of a separate geographic combatant command for Africa.

Congress also periodically includes provisions in annual National Defense Authorization Acts calling for DOD studies and reports on certain aspects of COCOM structure and operations.

These requirements, in addition to providing information to Congress, also serve the purpose of identifying areas of congressional concern which can influence DOD COCOM-related resourcing and policy decisions.

THE CURRENT UCP

The 2011 UCP is a classified document. On April 8, 2011, DOD released the 2011 UCP and the unclassified highlights were included in the following news release:

DOD Releases Unified Command Plan 2011[21]

The Department of Defense released today the updated Unified Command Plan (UCP), a key strategic document that establishes the missions, responsibilities, and geographic areas of responsibility for commanders of combatant commands. Unified Command Plan 2011, signed by the President on April 6, assigns several new missions to the combatant commanders.

Every two years, the chairman of the Joint Chiefs of Staff is required to review the missions, responsibilities, and geographical boundaries of each combatant command and recommend to the President, through the Secretary of Defense, any changes that may be necessary. As in past years, the 2011 review process included the combatant commanders, service chiefs and DOD leadership. Significant changes made by UCP 2011 include

- Shifting areas of responsibilities boundaries in the Arctic region to leverage long-standing relationships and improve unity of effort. As a result of this realignment, responsibility for the Arctic region is now shared between USEUCOM and USNORTHCOM rather than USEUCOM, USNORTHCOM and USPACOM as directed in previous UCPs.
- Giving USNORTHCOM responsibility to advocate for Arctic capabilities.
- Codifying the President's approval to disestablish U.S. Joint Forces Command.
- Expanding U.S. Strategic Command's responsibility for combating weapons of mass destruction and developing Global Missile Defense Concept of Operations.
- Giving U.S. Transportation Command responsibility for synchronizing planning of global distribution operations.

UCP 2011 continues to support U.S. defense security commitments around the world while improving military responsiveness to emerging crises.

A map with the UCP COCOM Areas of Responsibility is included in the Appendix.

Change One to the 2011 UCP[22]

On September 12, 2011, President Obama signed Change One to the 2011 UCP which primarily captured the administrative changes reflecting the disestablishment of USJFCOM and a number of SECDEF-directed Efficiency Initiatives. These changes include

- Removing any language referring to USJFCOM which was disestablished on August 31, 2011;
- Removing language for geographic combatant command standing joint force headquarters,[23] which are approved for disestablishment by the end of FY2012;
- Adding the responsibility of global standing joint headquarters to USTRANSCOM;
- Transferring the Joint Warfare Analysis Center missions from USJFCOM to USSTRATCOM; and

- Removing language and responsibilities for Information Operations, Military Deception, and Operations Security from USSTRATCOM as these mission areas are to be transferred to the Joint Staff.

Origins of the COCOMs

The nonstatutory origins of COCOMs are rooted in the U.S. experience in World War II. Prior to World War II, the Services operated independently and, despite lessons learned from World War I suggesting the Army and Navy needed to better communicate and plan, no real concerted effort was made to coordinate the armed forces, largely attributed to "bureaucratic distrust and service rivalry."[24] During this period, Marine Corps fears that the Army would lobby to eliminate the Marines on the grounds they were a "redundant service" as well as Army efforts to maintain control over the country's air arm typified the climate among the Services that made any meaningful reform virtually impossible.[25]

World War II presented unique challenges not faced during the eighteen month U.S. involvement in the First World War. While World War I was fought in a variety of theaters, such as Europe, Africa, the Mediterranean, and the Middle East, U.S. involvement was primarily limited to Europe and was predominately land-centric. In terms of strategic planning and command relationships, the United States played a supporting role.

The United States' experience in World War II bore little resemblance to that of the Great War. The European and Pacific theaters of the Second World War varied significantly, with the European Theater being a land-centric conflict supported by naval operations whereas the Pacific Theater was naval-centric and supported by Marine and Army ground forces. Both theaters also featured extensive supporting air force operations, including long-range strategic bombing campaigns unprecedented in both size and scope. In terms of relationships with allies, the United States assumed the leadership role in both the Pacific and European theaters - largely due to its unmatched military and industrial resources—despite insistence that the U.S. was "co-equal partners" with Great Britain, France, and Russia. Unlike 1918, after the Japanese surrender in 1945, U.S. political and military leaders did not advocate a post war policy of isolationism, because of fears of a communist Russia and, to a lesser extent, China. U.S. global military presence was viewed as a guarantee against unfettered communist expansion, and this presence necessitated an effective, geographically-focused, long term, joint command arrangement.

What Laws Govern COCOMs?

As previously noted, COCOMs are governed by the provisions contained in Sections 161 through 168 of Title X, Armed Forces, U.S. Code.[26] These sections address the following provisions;

- Section 161: The establishment of COCOMs;
- Section 162: Chain of command and assignment of forces for COCOMs;
- Section 163: Role of the CJCS;

- Section 164: Assignment and powers and duties of commanders of COCOMs;
- Section 165: Administration and support of COCOMs;
- Section 166: COCOM budget proposals ;
- Section 166a : Funding COCOMs through the CJCS ;
- Section 166b: Funding for combating terrorism readiness initiatives;
- Section 167: Unified COCOMs for special operations forces;
- Section 167a: Unified COCOMs for joint warfighting experimentation: acquisition authority; and
- Section 168: Military-to-military and comparable activities.

General Discussion of Provisions[27]

These provisions assign a number of responsibilities to the CJCS including a regular (at least every two years) review of the missions, responsibilities, areas of operation, and force structure of each combatant command. Upon completion of this review, the Chairman provides suggestions to the President—through the SECDEF—for changes in missions, force structure, and responsibilities for the COCOMs. These provisions also tie the Services to the COCOMs as the Secretaries of the military departments are directed to assign their forces (unless assigned elsewhere such as a multi-national peacekeeping operation) to COCOMs. These forces can only be transferred from the commands by the SECDEF.

Command Authority

Forces assigned to COCOMs are under the command of that COCOM commander, with the chain of command starting with the President and running through the SECDEF as indicated in *Figure 1.* The CJCS serves as a link between the President and the SECDEF and the COCOM commanders.

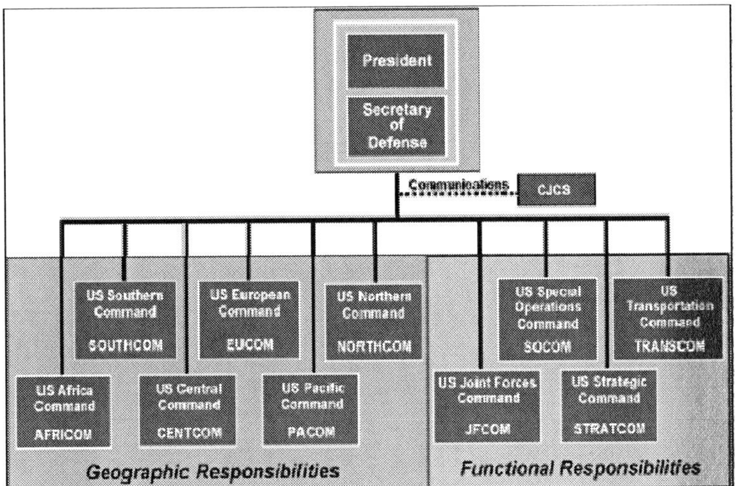

Source: This figure is taken from the author's National Defense University academic notes, 2010.
Note: USJFCOM was disestablished in August 2011 and no longer functions as a COCOM.

Figure 1. Combatant Command Chain of Command.

The President can send guidance to COCOM commanders through the CJCS, and the chairman can relay combatant commander's needs and concerns to the SECDEF and the President. The CJCS may exercise oversight of the COCOMs, if desired by the SECDEF, but has no command authority over the COCOMs.In this regard, the CJCS is described as taking part in national security discussions but not in the formal decision-making process as it relates to COCOMs.

COCOM Commander's Responsibilities

COCOM commanders are responsible for the accomplishment of missions assigned to them as well as all aspects of joint training, logistics, and military operations. COCOM commanders are also responsible for establishing command relationships with subordinate commands as well as organizing subordinate units as deemed necessary. While Combatant Commanders exercise control over subordinate units from different services, the services retain administrative control of their personnel to include assignment, promotion, schooling, and retirement. To facilitate administrative control, geographic combatant commands have service subcomponents for each service.

COCOM FUNDING

On an annual basis, COCOMs request Operations and Maintenance (O&M) funding. Funding for forces assigned to COCOMs are funded by the respective Services and funding for operations are funded separately, such as operations in Iraq and Afghanistan and counterterror operations which have primarily been funded through the Overseas Contingency Operations (OCO) account. *Table 1* provides O&M funding for the COCOMs from FY2009-FY2011. It should be noted that some amounts include operational and OCO costs (see corresponding notes).

COCOM Budgetary Provisions

The SECDEF is required to submit an annual budget proposal for the COCOMs and funding may be requested for joint exercises, force training, contingencies, and selected operations. Proposed funding for special operations forces (SOF) training with foreign forces may also be requested. COCOMs can also receive funds through the CJCS as part of the "Combatant Commander Initiative Fund."

Although not a COCOM, the U.S. element of the North American Aerospace Defense Command (NORAD) is also eligible for this fund. Authorized activities include:

- Force training;
- Contingencies;
- Selected operations;
- Command and control;
- Joint exercises to include activities of participating foreign nations;
- Humanitarian and civic assistance;

- Military education and training to military and related civilian personnel of foreign countries;
- Personnel expenses of defense personnel for bilateral or regional cooperation programs;
- Force protection; and
- Joint warfighting capabilities.

COCOMs also have access to a DOD budget account known as the "Combating Terrorism Readiness Initiatives Fund." Authorized activities under this fund include

- Procurement and maintenance of physical security equipment;
- Improvement of physical security sites;
- Under extraordinary circumstances:
- Physical security management planning;
- Procurement and support of security forces and security technicians;
- Security reviews, investigations, and vulnerability assessments; and
- Any other activity relating to physical security.

Table 1. FY2009-FY2011 O&M Budget (millions of dollars)

Program	FY2009[a] Actual	FY2010[a] Actual	FY2011[b] Estimate
U.S. Africa Command	213	306	296
U.S. Central Command	277	406	170
U.S. European Command	95	143	136
U.S. Joint Force	195	—	—
Command			
U.S. Northern Command	213	24	229
U.S. Pacific Command	266	278	290
U.S. Southern Command	163	188	87
U.S. Strategic Command	471	525	652
TOTAL(funded by the Services)	1,893	2,070	1,960
U.S. Special Operations Command[c]	6,240	6,712	3,944
U.S. Transportation Command[d]	12,083	14,231	13,493

Source: This information was provided to CRS by the Office of the Secretary of Defense in August 2011.

Notes: COCOM amounts reflect HQ and Mission Support O&M Funding. USJFCOM was disestablished in August 2011 and no longer functions as a COCOM.

[a] FY2009 and FY2010 includes OCO funding.

[b] FY2011 does not include OCO funding.

[c] SOCOM includes HQ and operational funds.

[d] TRANSCOM is funded predominately with customer orders.

FUNCTIONAL AND GEOGRAPHIC COMBATANT COMMANDS

Basic Organizational Principles[28]

COCOM commanders hold four-star flag rank and have risen through the ranks of their respective services, commanding at the highest levels. COCOM commanders have also met Joint Military Education Requirements as set forth in the Goldwater-Nichols Act. The President nominates combatant commanders based on the recommendations of the SECDEF. The Senate Armed Services Committee holds confirmation hearings for the nominees and the Senate then votes to confirm the candidates. While four-star officers from any service may serve as combatant commander for any given COCOM, some appointments (e.g., U.S. Pacific Command being commanded by a Navy admiral) traditionally have gone to specific services.

The basic configurations of COCOM staffs are generally the same and mirrors the Joint Staff at the Pentagon. COCOM staffs are organized as follows, although there are variations based on unique COCOM mission areas:

- J-1 Directorate of Manpower and Personnel;
- J-2 Directorate of Intelligence;
- J-3 Directorate of Operations;
- J-4 Directorate of Logistics;
- J-5 Directorate of Strategic Plans and Policy;
- J-6 Directorate of Command, Control, Communication, and Computer;
- J-7 Directorate of Operational Planning and Joint Force Development;
- J-8 Directorate of Force Structure, Resources, and Assessment; and
- J-9 Directorate of Interagency Partnering.

Within the COCOM command and staff construct, Joint Task Forces (JTFs) are often created to address a single policy concern and allocate resources, such as anti-drug efforts or humanitarian assistance, on a short to mid-term basis. JTFs can also be established in response to a crisis or for a long-term commitment.

Some COCOMs also have a political advisor (POLAD) assigned to the commander to serve as an interface with the civilian portion of the national security establishment as well as the ambassadors and embassy staffs of countries that fall under the COCOM commander's UCP mandate.

Interagency Representation in COCOMs[29]

Both Functional and Geographic COCOMs have integrated assets and representatives of other agencies and departments of the U.S. government into the COCOM's structure to enhance operations. Examples of this representation include:

- USAFRICOM: A State Department Deputy Commander for Civil-Military Activities, a senior U.S. Agency for International Development (USAID) advisor, and two other senior U.S. diplomats who serve as a Foreign Policy Advisor and as the J-9, Director of Outreach;

- USCENTCOM: An Interagency Action Group (IAG) established in the J-3 Directorate of Operations to integrate USCENTCOM and Interagency activities;
- USEUCOM: Established a J-9 Directorate for Interagency Partnering; and
- USNORTHCOM: A Joint Interagency Coordinating Group (JIACG) that integrates and synchronizes the activities of numerous civilian, State, Federal, and private sector organizations.

FUNCTIONAL COMBATANT COMMANDS
U.S. JOINT FORCES COMMAND (USJFCOM)
WEBSITE: HTTP://WWW.JFCOM.MIL/

Mission[30]

USJFCOM provided mission-ready joint-capable forces and supported the development and integration of joint, interagency, and multinational capabilities to meet the present and future operational needs of the joint force.

USJFCOM with its component commands responded to combatant commanders' operational requirements by providing timely and mission-ready joint capabilities. To accomplish this USJFCOM became a harmonized force made up of soldiers, sailors, airmen, Marines and civilians who excelled at joint warfighting. Together with the service headquarters and interagency and multi-national partners, USJFCOM achieved unity of effort to maximize joint strengths.

History[31]

From USJFCOM's Facebook page:

> USJFCOM was formed in 1999 when the old United States Atlantic Command (USLANTCOM) was renamed and given a new mission: leading the transformation of the U.S. military through experimentation and education. USLANTCOM had been active from 1947 to 1993 as a primarily U.S. Navy command, focused upon the wartime defense of the Atlantic sea lanes against Soviet attack. After the end of the Cold War, a 1993 reorganization gave the Command a new acronym, USACOM, and brought United States Army Forces Command and Air Combat Command under its authority.
>
> In late 2004, U.S. Joint Forces Command assumed the role of primary conventional force provider. This landmark change assigned nearly all U.S. conventional forces to Joint Forces Command.
>
> Requirements, for example, for U.S. service personnel to support the transformation of the Armed Forces of Liberia, were fed to JFCOM, in this case via Africa Command, and JFCOM liaised with the service staffs to obtain available forces. Along with this responsibility came the task to develop a new 'risk-assessment' process that provided national leaders a world-wide perspective on force-sourcing solutions.

USJFCOM Disestablished

On August 9, 2010, Secretary Of Defense Robert Gates announced a series of efficiency initiatives intend to reduce overhead, duplication, and excess within DOD.[32] One of Secretary of Defense Gates' recommendations was to disestablish USJFCOM, noting

> Over time, it has created an unneeded extra layer and step in the force management process. JFCOM's force management and sourcing functions will be assigned to the Joint Staff while the remaining responsibilities will be evaluated and those determined to be essential will be re-assigned to other entities.[33]

The President approved Secretary Gates' recommendation to close USJFCOM despite some Congressional opposition as well as from state and local officials concerned about lost jobs and the impact on the local economy. According the JFCOM commander, the closure was forecasted to save $400 million per year and result in the loss of about 1,900 local jobs.[34] On August 4, 2011, JFCOM was formally disestablished and a number of its functions transferred to the Joint Staff and other commands.

Subcomponents

The Service components of JFCOM—Fleet Forces Command, Air Combat Command, Marine Forces Command, and Army Forces Command—reverted to service control on August 1, 2011.[35] In addition, Special Operations Command, JFCOM was transferred to USSOCOM. JFCOM functions are planned to be realigned as follows:

Realignment of JFCOM Functions[36]

At the time of its disestablishment, JFCOM had 27 functional missions. Some of these mission areas were eliminated due to redundancy but others were deemed essential and aligned as described below.

- JFCOM's role as joint force provider will now be the responsibility of the Director, Joint Staff for Operations (J-3);
- JFCOM's Joint Capability Development and Integration (J-8) Directorate will merge with the Joint Systems Integration Center and Joint Fires Integration and Interoperability Team and these JFCOM organizations will be reassigned to the Director, Joint Staff Force Structure, Resources, and Assessments (J-8);
- JFCOM J-9 Joint Concept Development and Experimentation Directorate will be reassigned to the Director, Joint Staff Joint Coalition Warfare (J-9);
- The Joint Enabling Capabilities Command (JECC) is reassigned to USTRANSCOM;
- The Joint Warfare Analysis Center is reassigned to USSTRATCOM;
- The Joint Personnel Recovery Agency is reassigned to the Air Force; and
- The NATO School is reassigned to USEUCOM.

The physical move of all realigned JFCOM activities is expected to be completed by March 2012.

Selected Current Issues

With the disestablishment of USJFCOM and the reassignment of selected missions and functions, it remains to be seen if these changes will be effective and, in DOD terms, "seamless" in their application. While these changes were advertised by DOD as cost-effective, a retrospective analysis might be warranted to determine if these changes actually resulted in an overall cost savings or if these costs were simply transferred to the new owning agencies or COCOMs. The disestablishment and realignment of USJFCOM might also serve as both a model and precedent for future COCOM realignments or closures.

U.S. Special Operations Command (USSOCOM)[37] Website: www.socom.mil

Mission[38]

USSOCOM's primary mission is to organize, train, and equip special operations forces (SOF) and provides those forces to the Geographic Combatant Commanders under whose operational control they serve. USSOCOM also develops special operations strategy, doctrine, and procedures for the use of SOF and also develops and procures specialized, SOF-unique equipment for its assigned forces. USSOCOM is also responsible for synchronizing DOD planning against terrorists and their networks on a global basis. This particular aspect of USSOCOM's mission requires working extensively with other non-DOD U.S. government agencies, sometimes referred to as the Interagency.

History[39]

The 1980 Desert One tragedy and the 1983 loss of 237 Marines in Beirut, combined with the command and control problems experienced during Grenada in 1983 heightened apprehensions about DOD's ability to manage the Services, including special operations forces who were "owned" by their respective service.

By 1983, there was a small but growing sense in Congress of the need for military reforms. In June, the Senate Armed Services Committee (SASC), under the chairmanship of Senator Barry Goldwater (R-AZ), began a two-year-long study of DOD which included an examination of SOF. With concern mounting on Capitol Hill, DOD created the Joint Special Operations Agency on January 1, 1984. This agency had neither operational nor command authority over any SOF units and did little to address SOF issues.

Within Congress, there was a growing sense that a radical restructuring of SOF was needed. Proponents included Senators Sam Nunn (D-GA) and William Cohen (R-ME), both members of the SASC, and Representative Dan Daniel (D-VA), the chairman of the Readiness Subcommittee of the House Armed Services Committee (HASC). Congressman Daniel believed the U.S. military establishment had little interest in special operations and that U.S. SOF was second-rate when compared to countries such as Great Britain and Israel. Senators Nunn and Cohen also believed that DOD was not preparing adequately for future threats. Senator Nunn expressed a growing frustration with the Service's practice of reallocating monies appropriated for SOF modernization to non-SOF programs and suggested the U.S. needed a more efficient organization and a more direct chain of command for special operations.

In early 1986, SOF advocates introduced reform bills in the House and Senate. The Senate bill, co-sponsored by Senator Nunn and others called for a joint military organization for SOF and the establishment of an office in DOD to ensure adequate funding and policy emphasis for low-intensity conflict and special operations. Representative Daniel's proposal went even further - he wanted a national special operations agency headed by a civilian who would bypass the Joint Chiefs and report directly to the SECDEF , thereby keeping Joint Chiefs and Services out of the SOF budget process.

Congress held hearings on the two bills in the summer of 1986. CJCS Admiral William J. Crowe, led the Pentagon's opposition to the bills and proposed instead a new special operations forces command led by a three-star general. This proposal was not well received by Congress who wanted a four-star officer in charge so that he could deal on an equal footing with the four-star Service Chiefs.

President Reagan approved the establishment of USSOCOM on April 13, 1987. DOD activated USSOCOM on April 16, 1987 and nominated Army General Lindsay to be USSOCOM's first commander.

Subcomponents[40]

U.S. Army Special Operations Command (USASOC)

USASOC includes Army Special Forces, also known as Green Berets; Rangers; Civil Affairs, and Military Information Support Operations (MISO)—formerly known as psychological operations (PSYOPS)—units.

In addition, the 160th Special Operations Aviation Regiment (SOAR) provides rotary wing support to all SOF units.

Naval Special Warfare Command (NSWC)

NSWC consists of Sea, Air, and Land (SEAL) teams that conduct operations in both maritime and ground environments. NSWC also has SEAL Delivery Vehicle (SDV) teams—specialized SEALs that pilot small submersible vehicles that can deliver SEALs to their area of operations. NSWC includes Special Boat Teams that can deliver SEALs from ship to shore as well as operate in the littorals and rivers.

Air Force Special Operations Command (AFSOC)

AFSOC provides specialized fixed and rotary wing support to USSOCOM units. In addition to aircraft support, AFSOC also provides Combat Controllers, Pararescue Jumpers, Special Operations Weather Teams, and Tactical Air Control Parties (TACPs) to support special operations.

AFSOC is currently establishing a capacity to train and advise partner nation aviation units as part of foreign internal defense initiatives.

Marine Special Operations Command (MARSOC)

Established in 2005, MARSOC is the newest USSOCOM subcomponent. It consists of three Marine Special Operations Battalions, a Marine Special Operations Support Group, a Marine Special Operations Intelligence Battalion, and the Marine Special Operations School.

Joint Special Operations Command (JSOC)[41]

According to USSOCOM, JSOC is a sub unified command charged with studying special operations requirements and techniques to ensure interoperability and equipment standardization. JSOC also plans and conducts special operations exercises and training and develops joint special operations tactics.

USOCOM also notes JSOC "is comprised of an impressive amalgamation of rigorously screened and accessed, Soldiers, Sailors, Airmen, Marines, and Civilians," and "past and present members of JSOC have participated in all of our Nation's wars and contingency operations since it was activated in 1980." Press reports suggest JSOC is home to USSOCOM's national mission forceswhich reportedly conduct highly sensitive combat and supporting operations against terrorists on a world-wide basis.[42]

Other Components

Joint Special Operations University (JSOU)[43]

JSOU's stated mission is to

> Educate Special Operations Forces executive, senior, and intermediate leaders and selected other national and international security decision-makers, both military and civilian, through teaching, research, and outreach in the science and art of Joint Special Operations.

Special Operations Command–Joint Capabilities (SOC-JC)44

With the August 2010 disestablishment of JFCOM, SOCJFCOM was transferred to USSOCOM where it was renamed SOC-JC. SOC-JC's mission is to train conventional and special operations force commanders and their staffs in the employment of Special Operations Forces focusing on the full integration of SOF and the conventional forces in both planning and execution to enhance warfighting readiness.

Ongoing Operations

USSOCOM operates on a global basis in both overt and classified modes. Missions range from foreign internal defense to counterterrorism, but the primary emphasis for U.S. SOF is attacking terrorists and terror cells world wide.

The primary focus of these activities is the USCENTCOM region, where the USSOCOM commander testified in March 2011 that 85% of USSOCOM's deployed forces are operating.[45]

Selected Current Issues

Enablers

USSOCOM is seeking an increase in "enabling forces" units such as engineers, military police, medical, and maintenance that are provided by the Services but are not part of USSOCOM's organic force structure. These and other enabling forces provide support for forward-deployed SOF units in theaters such as Iraq and Afghanistan. The availability of and access to these enabling units will likely become increasingly important as conventional U.S. forces near completion of their withdrawal in Iraq and begin drawing down forces in Afghanistan in late 2011. U.S. SOF forces are expected to keep high levels of forces in the USCENTCOM AOR for the foreseeable future and many observers believe they will need enabling forces to support their operations.

"Fraying" of the Forces

The USSOCOM Commander testified in March 2011 that the decade-long wars had resulted in some "fraying around the edges" for U.S. SOF.[46] This almost constant state of deployment has resulted in significant time away from families and limited time for needed professional training and education.

Because USSOCOM's self-imposed annual growth rate of 4%—USSOCOM leaders argue growth in excess of 4% will dilute the quality of the force—adding SOF to alleviate these concerns is not possible. USSOCOM is exploring a variety of initiatives to address stress on the force.

Access to Local Training Areas

While USSOCOM has built a number of training ranges, they must also use pre-existing Service ranges and SOF does not always have priority for use and sometimes must travel great distances to use these facilities—contributing to "days away from home," a force stressor.

USSOCOM is in the process of building new ranges and coordinating access to existing service ranges, particularly those ranges that cannot be replicated due to environmental or land use restrictions.

U.S. Strategic Command (USSTRATCOM)
Website: http://www.stratcom.mil

Mission[47]

USSTRATCOM's primary responsibility is the stewardship and employment of U.S. nuclear weapons and to detect, deter, and prevent attacks against the United States and our allies and to join with the other combatant commands to defend the nation should deterrence fail.

Specific responsibilities include planning, synchronizing, advocating, and employing capabilities to meet the United States' strategic deterrence; space operations; cyberspace operations; global strike; missile defense; intelligence, surveillance, reconnaissance (ISR); and combating weapons of mass destruction (WMD).

History[48]

USSTRATCOM was established October 1, 2002. USSTRATCOM has provided intelligence, planning and cyber support to coalition forces in Afghanistan and Iraq. It monitors orbiting satellites and space debris, allowing spacecraft to avoid collision. USTRATCOM has also deployed systems to provide limited protection against ballistic missile attack.

The missions most directly associated with USSTRATCOM and its predecessors are deterrence and global strike. These were the missions of Strategic Air Command (SAC) from 1946 to 1992 and of the first USSTRATCOM from 1992 to 2002. Though best known for its connection with the nuclear deterrent, SAC conducted conventional bombing operations during the Korean War and Vietnam War and the first Persian Gulf War, 1991.

On June 1, 1992, SAC was replaced by a new unified command, USSTRATCOM. The new command's primary mission was to deter attack, especially nuclear attack, on the United States and its allies and, if deterrence failed, employ nuclear forces in response.

The U.S. military began operating in space in the late 1950s, with many of the early systems developed to meet SAC's needs for surveillance, warning, meteorology, and communications. By 1985, space activities had grown to such a scale that DOD created a new unified command, USSPACECOM, to manage military space operations. Secretary Rumsfeld's initiative to merge USSTRATCOM and USSPACECOM led to the creation of the current USSTRATCOM in 2002.

Two other areas took on increasing importance beginning around 2000: missile defense and cyberspace operations. By September 2004, the U.S. had deployed a limited system that offered some protection to North America and had opened discussions about extending the system to cover allies.

The U.S. military's reliance on computer networks grew exponentially in the 1980s and 1990s. National leaders took steps to protect defense networks in 1998, creating a Joint Task Force for Computer Network Defense and assigning it to USSPACECOM.

As computer attacks against DOD become more sophisticated and frequent there were calls to place greater emphasis and visibility on cyber operations.

Defense Secretary Robert Gates favored a new sub-unified command under USSTRATCOM that would recombine offensive and defensive computer network operations. Established 21 May 2010, U.S. Cyber Command (USCYBERCOM) was fully operational on Oct. 31, 2010.

Subcomponents

Service Subcomponents[49]

Air Force Global Strike Command (AFGSC)

AFGSC is responsible for the Air Force's three intercontinental ballistic missile (ICBM) wings, two B-52 Stratofortress wings and the sole B-2 Spirit wing. AFGSC has two numbered air forces that are tasked with providing capabilities to combatant commands. The

Eighth Air Force controls the long-range nuclear bomber assets (B-52s and B-2s) and the Twentieth Air Force controls the ICBM wings.

Air Force Space Command (AFSPC)

AFSPC provides space and cybersecurity forces for USSTRATCOM. It has two numbered air forces providing these capabilities. The Fourteenth Air Force controls and supports several satellite systems including the Global Positioning System (GPS); Defense Satellite Communications Systems Phase II and III; and the Defense Meteorological Support Program. In addition, the Fourteenth Air Force has Atlas, Delta, and Titan launch vehicles at its disposal to put payloads into orbit. The Twenty-Fourth Air Force plans and conducts cyberspace operations in support of combatant commands.

U.S. Army Forces Strategic Command (ARSTRAT)

ARSTRAT conducts space and missile defense operations and provides planning, integration, control, and coordination of Army forces and capabilities in support of USSTRATCOM missions.

Fleet Forces Command

Fleet Forces Command is responsible for the entire Atlantic Ocean, the Caribbean Sea, and the waters around Central and South America extending into the Pacific to the Galapagos Island.

Marine Corps Forces U.S. Strategic Command (MARFORSTRAT)

MARFORSTRAT serves as the Marine Corps service component to USSTRATCOM.

Functional Components[50]

U.S. Cyber Command (USCYBERCOM)

USCYBERCOM is a sub unified command that is subordinate to USSTRATCOM.

USCYBERCOM plans, coordinates, integrates, synchronizes, and conducts activities to defend DOD information networks and also conducts cyber space activities to enable U.S. military activities.

Joint Functional Component Command-Global Strike (JFCC-GS)

JFCC-GS optimizes planning, integration, execution and force management of assigned missions to deter attacks against the United States, its territories, possessions, and bases.

Joint Functional Component Command– Integrated Missile Defense (JFCC-IMD)

JFCC-IMD synchronizes operational-level global missile defense planning, operations support, and the development of missile defense effects for DOD.

Joint Functional Component Command-Intelligence, Surveillance and Reconnaissance (JFCC-ISR)

JFCC-ISR plans, integrates, and coordinates intelligence, surveillance and reconnaissance in support of strategic and global operations and strategic deterrence.

Joint Functional Component Command- Space (JFCC-Space)

JFCC-Space is responsible for executing continuous, integrated space operations to deliver theater and global effects in support of national and combatant commander objectives.

Joint Information Operations Warfare Center (JIOWC)

JIOWC provides joint information operations planning, execution, and operational-level integration of Electronic Warfare, Operations Security, and Military Deception across DOD to support USSTRATCOM, joint force commanders, and U.S. national objectives. On September 12, 2011, President Obama signed Change One to the 2011 UCP which transfers the Information Operations, Military Deception, and Operations Security missions from USSTRATCOM to the Joint Staff so it is possible the structure and missions of JIOWC may change significantly in the near future.

USSTRATCOM Center for Combating Weapons of Mass Destruction (SCC-WMD)

SCC-WMD plans, advocates, and advises the USSTRATCOM commander on WMD-related matters.

Ongoing Operations[51]

According to USSTRATCOM's commander, USSTRATCOM's major ongoing operation is to detect, deter and prevent attacks against the United States and to join with the other combatant commands to defend the nation should deterrence fail. One aspect of this operation is the "around the clock" command and control of U.S. nuclear forces. USSTRATCOM is also involved in implementing the new START treaty and efforts to sustain and modernize the nuclear triad and the nuclear weapons complex.

USSTRATCOM provides support to other combatant commanders in the areas of integrated missile defense and ISR operations. Not unlike its nuclear deterrence activities, USSTRATCOM and USCYBERCOM operate on a daily basis to improve their ability to operate and defend the DOD information network and make sure critical activities can continue, even in the face of adversary attempts to deny or disrupt them. USSTRATCOM is also responsible for U.S. military space operations on a day-to-day basis such as launching satellites and monitoring activities in space.

Selected Current Issues[52]

Funding the Nuclear Enterprise[53]

In March 2011, the USSTRATCOM commander noted during a Senate Armed Services Committee (SASC) that the FY2011 budget and FY2012 budget request represented a reversal of several years of a downward trend in nuclear enterprise funding. These funds are used for facilities, equipment, and personnel for sustaining and managing the nation's nuclear weapons and to dismantle weapons no longer needed. This sustained level of funding is viewed as essential to maintain adequate stockpile surveillance and assessment programs and synchronize weapons sustainment programs by introducing cost effective safety and security features for legacy nuclear weapons.

***Support for a Fledgling USCYBERCOM*[54]**

The USSTRATCOM commander has expressed a requirement to accelerate the acquisition of "comprehensive, shared cyber awareness tools to expand opportunities to secure critical information, reduce points of vulnerability, and develop responses to ensure warfighter access to essential information systems."[55] The urgency of this requirement is likely underscored by the increasing tempo and effectiveness of cyber attacks against DOD information systems. While access to cyber tools was cited as a command priority, recruiting adequately trained cyber professionals is proving to be a challenge, and USCYBERCOM is developing career paths to develop and retain cyber professionals. One means to develop and sustain CYBERCOM's workforce is to partner with educational and commercial informational technology entities.

New Strategic Command and Control Complex and Nuclear Command, Control, and Communication Node at Offut Air Force Base, Nebraska

The FY2012 budget request seeks a first increment of $150 million to begin replacing the current facilities housing these functions. The current complex was developed before the information technology revolution and, in addition to aging and eroding infrastructure, it no longer has the capacity to support current mission demands. The new facility is viewed as essential to meet current and future operational demands, and both continuity in funding as well a required level of funding will likely continue to be command concerns until the project is completed.

U.S. Transportation Command (USTRANSCOM)
Website: http//:www.transcom.mil

Mission[56]

Develop and direct the Joint Deployment and Distribution Enterprise to globally project strategic national security capabilities; accurately sense the operating environment; provide end-to-end distribution process visibility; and supportive of joint, U.S. government and Secretary of Defense-approved multinational and nongovernmental logistical requirements.

History[57]

World War II, the Berlin blockade, the Korean War, and the Vietnam War demonstrated the need for the United States to maintain a capable and ready transportation system for national security. A 1978 exercise exposed significant gaps in understanding between military and civilian participants: mobilization and deployment plans fell apart, and as a result, the United States and its NATO allies "lost the war."

Two major recommendations came out of the exercise. First, the Transportation Operating Agencies (later called the Transportation Component Commands) should have a direct reporting chain to the JCS. Second, the JCS should establish a single manager for deployment and execution. As a result, the JCS formed the Joint Deployment Agency (JDA) in 1979 at MacDill Air Force Base, FL.

Although the JDA had responsibility for integrating deployment procedures, it did not have authority to direct the Transportation Operating Agencies or Unified and Specified Commanders to take corrective actions, keep data bases current, or adhere to milestones. In April 1987 President Reagan ordered the SECDEF to establish a Unified Transportation Command (UTC), a directive made possible in part by the Goldwater-Nichols DOD Reorganization Act of 1986, which revoked the law prohibiting consolidation of military transportation functions.

Designated the United States Transportation Command (USTRANSCOM), its mission was to provide global air, sea, and land transportation to meet national security needs. It had three transportation component commands—the Air Force's Military Airlift Command (replaced by Air Mobility Command in 1992), the Navy's Military Sealift Command, and the Army's Military Traffic Management Command, (renamed Military Surface Deployment and Distribution Command in 2004). On June 22 1987, the President nominated Air Force General Duane H. Cassidy as the first USTRANSCOM commander. The commander of USTRANSCOM received operational direction from the National Command Authority (NCA) through the CJCS. There were, however, some deficiencies in this new command arrangement. The Services retained their single-manager charters for their respective transportation modes. Even more restrictive, USTRANSCOM's authorities were limited primarily to wartime.

As a result, during peacetime, USTRANSCOM's component commands continued to operate day-to-day much as they had in the past. They controlled their industrial funds and maintained responsibility for Service-unique missions, Service-oriented procurement and maintenance scheduling, and DOD charters during peacetime single-manager transportation operations. They also continued to have operational control of forces.

DOD learned much from the strategic deployment for Desert Shield/Desert Storm and foremost among those lessons was USTRANSCOM and its component commands needed to operate in peacetime as they would in wartime. Consequently, on February 14, 1992, the SECDEF gave USTRANSCOM a new charter. Stating the command's mission to be "to provide air, land, and sea transportation for DOD, both in time of peace and time of war," the charter greatly expanded the authorities of the USTRANSCOM commander. Under the new charter, the Service Secretaries assigned the components to the USTRANSCOM commander in peace and war. In addition, the military departments assigned to him, under his combatant command, all transportation assets except those that were Service-unique or theater-assigned. The charter also made the USTRANSCOM commander DOD's single-manager for transportation, other than Service-unique and theater-assigned assets.

On September 16, 2003, Secretary of Defense Rumsfeld designated the USTRANSCOM commander as the Distribution Process Owner (DPO) to serve "as the single entity to direct and supervise execution of the Strategic Distribution system" in order to "improve the overall efficiency and interoperability of distribution related activities - deployment, sustainment and redeployment support during peace and war." As the DPO, USTRANSCOM partnered with other COCOMs, the Services, defense agencies, Office of the Secretary of Defense, Joint Staff and industry to improve the Joint Deployment and Distribution Enterprise Since 2003, USTRANSCOM has gained additional responsibilities related to its role as the Distribution Process Owner. In 2004, USTRANSCOM became the portfolio manager for DOD logistics information technology systems, and received acquisition authority for procuring information technology systems, carrying out research projects and obtaining services needed to transform the DOD supply chain.

Subcomponents[58]

Surface Deployment and Distribution Command (SDDC)

"SDDC provides ocean terminal, commercial ocean liner service and traffic management services to deploy, sustain and redeploy U.S. forces on a global basis. The command is responsible for surface transportation and is the interface between DOD shippers and the commercial transportation carrier industry. This includes movement of servicemembers household goods and privately owned vehicles. SDDC is the nation's largest customer to the moving industry with more than 500,000 household goods moves a year. The command also provides transportation for troops and materiel to ports of departure in the U.S. and overseas and manages 24 ports worldwide, including military terminals at Sunny Point, N.C., and Concord, Calif."

Military Sealift Command (MSC)

"MSC provides sealift transportation services to deploy, sustain and redeploy U.S. forces around the globe. MSC provides sealift with a fleet of government-owned and chartered U.S.-flagged ships. MSC executes Voluntary Intermodal Sealift Agreement (VISA) contracts for chartered vessels. Sealift ships principally move unit equipment from the U.S. to theaters of operation all over the world.

In addition to sealift ships, MSC operates a fleet of prepositioned ships strategically placed around the world and loaded with equipment and supplies to sustain Army, Navy, Marine Corps, Air Force and Defense Logistics Agency operations. These ships remain at sea, ready to deploy on short notice, which significantly reduces the response time for the delivery of urgently needed equipment and supplies to a theater of operation."

Air Mobility Command (AMC)

"AMC provides strategic and tactical airlift, air refueling, and aeromedical evacuation services for deploying, sustaining and redeploying U.S. forces wherever they are needed. Many special duty and operational support aircraft are also assigned to AMC (including Air Force One).

In addition, AMC contracts with commercial air carriers through Civil Reserve Air Fleet (CRAF) and other programs for movement of DOD passengers and cargo. AMC's air fleet provides swift response as an element of America's global reach."

Ongoing Operations[59]

USTRANSCOM conducts military and commercial transportation, distribution process integration, terminal management, aerial refueling, and global patient movement on a daily basis. While USTRANSCOM supports all combatant commands, its largest customer is CENTCOM due to the conflicts in Iraq and Afghanistan.

In 2010, USTRANSCOM deployed and redeployed 48 Brigade Combat Teams, 75,000 Aerial Expeditionary Forces, 12 Security Force Packages, and also moved Marine Expeditionary, Stryker, and Combat Aviation Brigades in the USCENTCOM region. The centerpiece of these efforts was the on-time deployment of 30,000 surge troops to Afghanistan and the drawdown from 130,000 to 50,000 service members in Iraq by the end of August 2011.

Selected Current Issues[60]

Ground Supply in Afghanistan

USTRANSCOM continues efforts to expand surface networks which supply Afghanistan. Called the Northern Distribution Network (NDN), USTRANSCOM's stated priority is to enhance and improve this network.

In 2010, two additional routes through the Baltics and Central Asia were added which has facilitated faster and less costly cargo flow. The improvement of the NDN will take on added importance as the United States and NATO allies begin to withdraw significant numbers of troops and quantities of equipment over the next few years. Even after the withdrawal, maintaining the NDN to supply the Afghan military will also likely remain a priority.

New UCP Mission: "Global Distribution Synchronizer"

USTRANSCOM is slated to assume the responsibility as "global distributions synchronizer" as stipulated in the 2011 UCP.

This new mission will require USTRANSCOM to "assess the global distribution environment, identify potential distribution gaps, threats, seams, and vulnerabilities and develop distribution strategies that will ensure continued warfighter success."[61] This significant undertaking will likely involve extensive coordination and planning with all COCOMs and will eventually make USTRANSCOM responsible for the global movement of everything from "beans to bullets." [62]

Maintaining Air Mobility Readiness

In testimony to the SASC in April 2011, the USTRANSCOM commander testified that the recapitalization of the aerial tanker refueling fleet is his top acquisition priority. The new aerial refueling tanker presently under development, the KC-46A, is viewed as essential to support USTRANSCOM's aerial mobility missions.

He further noted that schedule delays or funding cuts to this program could have an adverse impact on aerial mobility as many of the aircraft they are intended to replace are near obsolescence.

Geographic Combatant Commands

U.S. Africa Command (USAFRICOM)[63]
Website: http://www.africom.mil

Mission

"Africa Command protects and defends the national security interests of the United States by strengthening the defense capabilities of African states and regional organizations and, when directed, conducts military operations, in order to deter and defeat transnational threats and to provide a security environment conducive to good governance and development."[64]

"USAFRICOM is responsible for U.S. military relations with 54 African countries including the islands of Cape Verde, Equatorial Guinea, and Sao Tome and Principe, along with the Indian Ocean islands of Comoros, Madagascar, Mauritius, and Seychelles. U.S. Central Command maintains its traditional relationship with Egypt, though USAFRICOM coordinates with Egypt on issues relating to Africa security."[65]

A Different Kind of Combatant Command[66]

Dr. Cynthia Watson, a professor at the National War College and author of "Combatant Commands: Origins, Structure, and Engagements," observes USAFRICOM has a different type of mission, noting

> USAFRICOM allows the U.S. government, particularly DOD, to work toward a more stable environment in which political and economic growth can take place with three former U.S. commands consolidated into a single new one to best work out the U.S. government efforts. Africa Command hopes to avoid that traditional combatant command goals of warfighting in favor of war prevention, making its orientation quite different from other parallel organizations.

History[67]

Africa's previous command arrangements [USEUCOM has had responsibility for most of Africa since the end of World War II][68] reflect the relatively low level of importance assigned to the African continent within the U.S. military structure. Before the creation of USAFRICOM, Africa generally received less attention than other regions under the three aforementioned military commands. USCENTCOM was focused on U.S. security priorities in Iraq and Afghanistan. USEUCOM was preoccupied with NATO, relations with European allies and Russia. USPACOM was primarily focused on regional powers such as China, India and North Korea. There was a consensus that the previous command arrangements for Africa represented a "suboptimal organizational structure." Secretary of Defense Robert Gates observed that previous command arrangements for the African continent were "an outdated arrangement left over from the Cold War." U.S. foreign policy in the region, like its military involvement, was primarily concerned with Cold War geopolitics rather than African policy and development. Scholars have referred to this policy attitude as "benign neglect," designating Africa as the "stepchild" of U.S. foreign policy. This attitude continued after the end of the Cold War, when the opportunity to articulate coherent policy was overlooked.

While the transformation of the geopolitical landscape was significantly altered by the fall of the Soviet Union, America's attention was focused on the newly freed Eurasian states. In 1995, in its U.S. Security Strategy for Sub-Saharan Africa, DOD noted that "ultimately we see very little traditional strategic interest in Africa."

Events in the late 1990s began to change U.S. perception of security interests in Africa. In 1998, two U.S. embassies in Africa were bombed. While many scholars believe these twin bombings marked a turning point in U.S. strategic policy toward the region, the domestic terrorist attacks of September 11, 2001, forced a reassessment of U.S. policy vis-à-vis Africa and its role in the global war on terrorism. The growing strategic importance of Africa for the United States was subsequently articulated in government documents. In the 2002 National Security Strategy, the concept of weak states and their role in global instability was an important theme. The 2006 National Security Strategy solidified the newly important role of Africa, observing, "Africa holds growing geo-strategic importance and is a high priority of this administration." The National Security Presidential Directive (NSPD 50) signed by President Bush in September 2006 provided the first update to overall U.S. strategy toward Africa since 1992.

Africa's abundance of natural energy resources has made it an attractive region for countries the United States, China, and India seeking additional resources. U.S. officials note three areas are of particular concern: (1) the number of soft targets (e.g., embassies and consulates); (2) the recruiting potential for young, angry, marginalized youth from Somalia to Morocco; and (3) the potential of sanctuary for international terrorists (particularly in large ungoverned spaces). Africa has also been a target terrorist activity: there were attacks on the U.S. embassies in Tanzania, and Kenya, in 1998; on targets in Mombasa, Kenya, in 2002; and in Algiers in 2007. In particular, the Horn of Africa is of concern to terrorism experts and military personnel.[69] In addition to terrorism concern, the growth of international piracy in the region has become a serious problem for the international community.[70]

On February 6, 2007, the White House announced a presidential directive to create a new unified combatant command in Africa. U.S. Africa Command (USAFRICOM) commenced official operations on October 1, 2007, and remained a sub-unified command under U.S. European Command (USEUCOM) until October 2008. On October 1, 2008, USAFRICOM was declared fully mission capable and took over the role of geographic combatant command for Africa. While the USAFRICOM headquarters remains in Germany, there are aspirations to possibly move it to Africa sometime in the future but regional sensitivities and security concerns have made this a challenging proposition.

Subcomponents

U.S. Army Africa (USARAF)[71]

Headquartered in Vicenza, Italy, USARAF is the Army component to USAFRICOM. In concert with national and international partners, it conducts sustained security engagement with African land forces to promote peace, stability, and security in Africa. If required, USAFRICOM can deploy as a contingency headquarters in support of crisis response. USAFRICOM is staffed by the Southern European Task Force (SETAF), which prior to August 2008 was an airborne task force that supported NATO in both combat and humanitarian operations.

U.S. Naval Forces, Africa (NAVAF)[72]

"NAVAF is part of a combined U.S. Naval Forces Europe (NAVEUR)/NAVAF headquarters and is tasked with the conduct of the full range of maritime operations and theater security cooperation in concert with coalition joint interagency and other partners in order to advance security and stability in Europe and Africa. Their combined areas of responsibility cover approximately half of the Atlantic Ocean, from the North Pole to Antarctica; as well as the Adriatic, Baltic, Barents, Black, Caspian, Mediterranean and North Seas. NAVEUR-NAVAF [headquartered in Naples, Italy] covers all of Russia, Europe and nearly the entire continent of Africa. It encompasses 105 countries with a combined population of more than one billion people and includes a landmass extending more than 14 million square miles." The U.S. Sixth Fleet supports NAVAF operations in the AFRICOM AOR and is headquartered in Naples, Italy.

U.S. Air Forces, Africa (AFAFRICA)[73]

"AFAFRICA, or 17th Air Forces, conducts sustained security engagement and operations to promote air safety, security, and development in Africa. AFAFRICA is located at Ramstein Air Base in Germany."

U.S. Marine Corps Forces, Africa (MARFORAF)

MARFORAF, located Stuttgart, Germany, is the Marine's service component headquarters for USAFRICOM. "MARFORAF conducts operations, exercises, training, and security cooperation activities throughout the African Continent."[74]

Combined Joint Task Force-Horn of Africa (CJTF-HOA)

"CJTF-HOA is USAFRICOM's forward operating task force located at Camp Lemonnier in Djibouti. It is one of more than 23 tenant organizations. CJTF-HOA has approximately 2,000 people assigned on rotating tours. While the core staff works at Camp Lemonnier, most of the service members assigned to CJTF-HOA are "embedded" in partner nations performing a range of activities—building partner security capability, capacity, and infrastructure through regional cooperation; military-to-military programs; civil-military affairs projects; and professional military education programs. Through an indirect approach, the task force, along with coalition and other U.S. defense components, provides support to regional organizations to help foster cooperation, enhance collective peace-keeping, improve humanitarian assistance, and support civil-military operations."[75]

Special Operations Command-Africa (SOCAFRICA)

"On October 1, 2008, SOCAFRICA was established as USAFRICOM's Theater Special Operations Command—a functional, sub-unified special operations command for Africa. SOCAFRICA contributes to USAFRICOM's mission through the application of the full spectrum of special operations forces capabilities including civil affairs, information operations, theater security cooperation, crisis response, and campaign planning."[76] SOCAFRICA is headquartered in Stuttgart, Germany.

Ongoing Operations

NATO Operations Against Libya[77]

On March 19, 2011, U.S. and coalition aircraft began air strikes in support of United Nations Security Council Resolution (UNSCR) 1973 to protect the Libyan people against Libyan armed forces loyal to the Qadhafi regime.78 USAFRICOM was placed in command of this operation (Operation Odyssey Dawn). On March 31st, NATO assumed command of this operation (renamed Operation Unified Protector) from USAFRICOM. The NATO mission ended on October 31, 2011, seven months after it began. NATO's current plan is to function in an "oversight role to make sure that the situation remains moving forward in a positive direction, with no threats or attacks on civilians."79 With the death of Qadhafi and the transition to a new regime, it is unclear what role USAFRICOM will play.

U.S. Deployment to Central Africa[80]

On October 14, 2011, President Obama informed Congress that on October 12, 2011,a small team of U.S. military personnel began deploying to Uganda and that by November about 100 U.S. military personnel—primarily U.S. Special Forces—would be deployed to Central Africa to act as advisors to partner forces who are attempting to kill or capture the leadership of the Lord's Resistance Army (LRA). The LRA is a nonreligious terror group that routinely kidnaps children and forces them to serve as soldiers which has committed multiple acts of terror in the region over the past two decades. U.S. forces will operate in Uganda, South Sudan, the Central African republic, and the Democratic Republic of the Congo, and the mission has been characterized as a mission of "months as opposed to years."

In the past, USAFRICOM has provided training and equipment to a variety of central African militaries. USAFRICOM's involvement has been credited with helping central African forces attrite the LRA to about 200 core fighters and about 600 supporters. It is not known what USAFRICOM's direct role in the current operation will be but it is possible that SOCAFRICA will exercise a degree of control and oversight of this mission.

Operation Enduring Freedom-Trans Sahara (OEF-TS)

"OEF-TS is the U.S. Military regional plan against terrorist and violent extremists. OEF-TS is the military component to Trans Sahara Counterterrorism Partnership (TSCTP), together with other USG organizations, OEF-TS will help enhance stability and deter terrorist activity on the African continent, with an emphasis on greater security in North Africa. USAFRICOM is working with international partners in a regional approach to common areas of concern such as commerce, education and economic development. OEF-TS works with the partner nations to expand military -to-military cooperation, ensure adequate resources are available to train, advise, assist regional units, and establish mechanisms to promote better regional cooperation, communications, and intelligence-sharing."[81]

Exercises[82]

USAFRICOM planned 13 major joint exercises in Africa for 2011. USAFRICOM uses these exercises to encourage the development of partner nation's security capabilities and instilling professionalism in Africa's various military and security forces. These exercises range from traditional land combat operations, to logistics and medical operations, humanitarian aid and disaster response, to counterterrorism training.

Selected Current Issues

Libyan Endstate

With the end of the Qadhafi regime, it is possible that USAFRICOM could take on a greater role in Libya. Potential operations include stability operations and security force assistance as any successor to the Qadhafi regime will likely require assistance in both areas. On a different level, USAFRICOM and partner nations might also assist Libya in re-establishing national-level military command and control such as a Ministry of Defense as well as service-centric command and control for Libyan land, air, and naval forces. As such, USAFRICOM could play an important long-term regional role in these and other related endeavors.

Transnational Threats[83]

The USAFRICOM commander has testified that violent extremism, piracy, and narcotics trafficking constitute three of the most dangerous threats in Africa. Somalia has been termed a failed state and linked to this instability, Al Qaeda has gained significant influence in east Africa over the past year. Piracy, once concentrated off the coast of Somalia, has spread as far south as the Mozambique Channel and east into the Indian Ocean despite increased international naval presence. Illicit trafficking of narcotics also poses a significant threat to regional stability in both West and East Africa. In and of themselves, each of these threats require complex, multi-national, political-military solutions and appropriate levels of resources. In addition to financial resources, USAFRICOM notes "maintaining a predictable pattern of available operational forces for sustained engagement activity is critical to the success of our efforts on the continent and an ongoing challenge."[84]

Exercise Funding[85]

As previously noted, due to Africa's colonial history and reluctance to align themselves with outside nations, military exercises play a central role in USAFRICOM's ability to function effectively in the region—perhaps more so than any other combatant command. A critical concern for many is in an era of budgetary uncertainty, USAFRICOM's various exercise funding sources will be subject to significant reductions, thereby adversely impacting AFRICOM's sustained engagement on the continent.

U.S. Central Command (USCENTCOM)
Website: http://www.centcom.mil

Mission

"With national and international partners, U.S. Central Command promotes cooperation among nations, responds to crises, and deters or defeats state and nonstate aggression, and supports development and, when necessary, reconstruction in order to establish the conditions for regional security, stability, and prosperity."[86]

History

The Iranian hostage crisis that played out from 1979 to 1981 and the 1979 Soviet invasion of Afghanistan served to underscore the need to strengthen U.S. presence in the region, President Carter established the Rapid Deployment Joint Task Force (RDJTF) in March 1980 and President Reagan took steps to transform the RDJTF into a permanent unified command over a two-year period. USCENTCOM was formally established on January 1, 1983.

By late 1988, the regional strategy largely focused on the potential threat of a Soviet invasion of Iran. The new USCENTCOM commander, General H. Norman Schwarzkopf, believed that an invasion of Iran was unlikely and began to focus on the possible emergence of a new regional threat—Iraq. On August 2, 1990, these beliefs became a reality when Iraq invaded Kuwait. The U.S. and other nations responded quickly by forming a coalition and deploying forces to Saudi Arabia to deter further Iraqi aggression. On January 17, 1991, U.S. and coalition forces launched Operation Desert Storm with an air interdiction campaign, which prepared the theater for a coalition ground assault. The primary coalition objective, the liberation of Kuwait, was achieved on February 27, and the next morning a cease-fire was declared, just one hundred hours after the commencement of the ground campaign.

Even though formal hostilities ended after the hundred hour war, there were other security concerns. Operation Provide Comfort was implemented in April 1991 to provide humanitarian assistance to the Kurds and enforce a "no-fly" zone in Iraq. In August 1992, Operation Southern Watch began in response to Saddam's noncompliance with U.N. Security Council Resolution 688 condemning his brutal repression of Iraqi civilians in southeastern Iraq. In January 1997, Operation Northern Watch replaced Provide Comfort and focused on enforcing the northern no-fly zone. Throughout the decade, USCENTCOM operations such as Vigilant Warrior, Vigilant Sentinel, Desert Strike, Desert Thunder (I and II), and Desert Fox responded to Iraqi threats to its neighbors or to enforce U.N. Security Council resolutions.

To prevent widespread starvation attributed to clan warfare, USCENTCOM undertook Operation Provide Relief in 1992 to supply humanitarian assistance to Somalia and northeastern Kenya as sanctioned by the U.N. In 1993, despite of some U.N. success in the countryside, the situation in Mogadishu worsened, and a series of violent confrontations compelled President Clinton to order the withdrawal of U.S. troops from Somalia. Throughout the decade following the Gulf War, terrorist attacks had a major impact on USCENTCOM forces in the region. In 1996, the Khobar Towers in Saudi Arabia were bombed, killing 19 U.S. airmen. In 1998 terrorists attacked the U.S. embassies in Kenya and Tanzania, killing 250 persons, including 12 Americans. The October 2000 attack on the USS Cole, resulting in the deaths of 17 U.S. sailors, was linked to Osama bin Laden's al Qaida organization.

The September 11, 2001, attacks compelled the U.S. to initiate a war against international terrorism. USCENTCOM launched Operation Enduring Freedom in October 2001 to expel the Taliban government in Afghanistan, which was harboring al Qaida terrorists, hosting terrorist training camps, and repressing the Afghan population. In the wake of 9/11, some members of international community found Iraq's lack of cooperation with United Nation Security Council (UNSC) Resolutions regarding weapons of mass destruction unacceptable. Continued Iraqi resistance led the UNSC to authorize the use of force by a U.S.-led coalition. Operation Iraqi Freedom began March 19, 2003.

Following the defeat of the Taliban regime in Afghanistan (November 9, 2001) and Saddam Hussein's government in Iraq (April 8, 2003), USCENTCOM continued to provide security to the new governments in those countries, conducting counterinsurgency operations and assisting host nation security forces to provide for their own defense.

Beginning in October 2002, USCENTCOM conducted operations in the Horn of Africa to assist host nations there to combat terrorism, establish a secure environment, and foster regional stability. USCENTCOM also provided disaster relief such as the October 2005 earthquake in Pakistan and the large-scale evacuation of American citizens from Lebanon in 2006.

On October 1, 2008, DOD transferred responsibility for Sudan, Eritrea, Ethiopia, Djibouti, Kenya, and Somalia to the newly established USAFRICOM while Egypt remained in the USCENTCOM AOR.[87]

Subcomponents

U.S. Army Central (ARCENT)[88]

ARCENT, in addition to being the Army component, has the mission to serve as an expeditionary headquarters to handle operations across the full spectrum for limited duration operations. The U.S. Third Army forms the basis of this subcomponent command and also serves as the Coalition Forces Land Component Command. ARCENT has a forward headquarters in Doha, Qatar.

U.S. Naval Forces Central Command (NAVCENT)[89]

NAVCENT has its headquarters in Manama, Bahrain, the homeport of the U.S. Fifth Fleet. NAVCENT forces in the region normally include either a Carrier Strike Group, Expeditionary Strike Group, or an Amphibious Strike Group. NAVCENT also serves as the command element for the Combined Maritime Forces, which is comprised of naval forces from about two dozen nations that are responsible for combating terrorism, piracy, and illegal drug trafficking in the region.

U.S. Air Forces Central (AFCENT)[90]

"Located at Shaw Air Force Base, South Carolina, the 9th Air Force is the headquarters for AFCENT and serves as the air component for a 27-nation area within the USCENTCOM AOR. The 9th Air Force is also an intermediate headquarters under Air Combat Command and is responsible for five active-duty flying wings, as well as overseeing the operational readiness of 18 designated units of the Air National Guard and Air Force Reserve."

U.S. Marine Corps Forces Central Command (MARCENT)[91]

"MARCENT is designated as the Marine Corps service component for USCENTCOM. MARCENT is responsible for all Marine Corps forces in the CENTCOM AOR. MARCENT provides Marine Expeditionary Forces capable of conducting a wide range of operations, offering the command a responsive and unique set of capabilities." MARCENT has a forward headquarters in Bahrain.

Special Operations Command Central (SOCCENT)[92]

"SOCCENT is headquartered at MacDill Air Force Base, FL and is a subordinate unified command of U.S. Central Command. It is responsible for planning Special Operations throughout the USCENTCOM AOR, planning and conducting peacetime joint/combined Special Operations training exercises and orchestrating command and control of peacetime and wartime Special Operations as directed. SOCCENT exercises operational control of assigned and attached SOF that deploy for the execution of training and for operational missions in the USCENTCOM area of operations as directed by the USCENTCOM commander. When directed by the USCENTCOM commander, SOCCENT forms a Joint Special Operations Task Force (JSOTF)."

Ongoing Operations[93]

USCENTCOM forces are conducting a theater-wide campaign in conjunction with other partner nations against Al Qaeda and its extremist allies. USCENTCOM's stated main effort is in Afghanistan where U.S., NATO, and coalition allies are conducting a counterinsurgency campaign as well as training, equipping and advising Afghan military and police forces so they can eventually take over security responsibilities for their country. In Pakistan, USCENTCOM is supporting Pakistan's military efforts against extremists operating from Pakistan that threaten both Pakistan and Afghanistan as well as supporting the U.S. government's overall strategic partnership with Pakistan. In Iraq, as U.S. forces are drawing down and security responsibilities are being transitioned to Iraq, USCENTCOM is standing up the Office of Security Cooperation– Iraq to conduct long-term security assistance and cooperation activities with Iraq.[94]

Selected Current Issues[95]

Middle East Unrest

Concern has been expressed over past and current unrest in Egypt, Bahrain, Jordan, Syria, and Yemen spurred by people's demands for democratic rights in those countries. In the case of Egypt, a long standing regime was deposed and according to some analysts the country is now moving to a more democratic form of government. Of critical concern to many is, despite overtures towards democracy, widespread demonstrations have resulted in varying degrees of instability in these countries. In the case of Egypt, Jordan, Yemen, and Bahrain, they are considered crucial counterterrorism partners and long term unrest and possible political changes could have a detrimental impact on U.S. counterterrorism efforts in the region. Furthermore, civil unrest or possibly civil war in any of these countries could place further demands on USCENTCOM if the U.S. government determines post-conflict security assistance is required.

Iranian Interference

U.S. intelligence agencies and USCENTCOM have long noted that Iran has provided arms, ammunition, money, and improvised explosive device (IED) components to insurgents opposing U.S. efforts in Iraq and Afghanistan. The continuation of this support undermines USCENTCOM efforts in the region, and concerns exist that after U.S. forces depart Iraq and eventually Afghanistan, Iran might step up its support, thereby furthering the security challenges faced by Iraqi and Afghan security forces.

Iraqi Security Force Needs after U.S. Departure

The USCENTCOM commander expressed concern after U.S. forces depart Iraq, logistics, maintenance, intelligence fusion, and the building of an Iraqi air force would continue to be challenges over the long term.

In his opinion, if left to their own, Iraqis would have a difficult time achieving these crucial requirements for an effective military. Given this state of affairs, it is possible some type of USCENTCOM or U.S. government program might be required to complete these important security tasks.

U.S. European Command (USEUCOM) Website: http://www.eucom.mil

Mission[96]

The mission of USEUCOM is to conduct military operations, international military partnering, and interagency partnering to enhance transatlantic security and defend the United States . USEUCOM forces constitute the United States military contribution to NATO.

USEUCOM Commander and NATO

The USEUCOM Commander also traditionally serves as the Supreme Allied Commander of NATO (SACEUR). SACEUR's responsibilities are outlined as follows:[97]

> SACEUR is responsible for the overall command of NATO military operations. He conducts the necessary military planning for operations, including the identification of forces required for the mission and requests these forces from NATO countries, as authorized by the North Atlantic Council and as directed by NATO's Military Committee. SACEUR analyzes these operational needs in cooperation with the Supreme Allied Commander Transformation.
>
> SACEUR makes recommendations to NATO's political and military authorities on any military matter that may affect his ability to carry out his responsibilities. For day-to-day business, he reports to the Military Committee, composed of Military Representatives for Chiefs of Defense of NATO member countries. He also has direct access to the Chiefs of Defense and may communicate with appropriate national authorities, as necessary, to facilitate the accomplishment of his tasks.

In the case of an aggression against a NATO member state, SACEUR, as Supreme Commander, is responsible for executing all military measures within his capability and authority to preserve or restore the security of Alliance territory.

History[98]

USEUCOM was established August 1, 1952 to provide unified command and authority over all U.S. forces in Europe. For several years after World War II the services had maintained separate commands in Europe that reported directly to the JCS. After the end of the occupation of Germany in 1949, some questioned the U.S. commitment to the defense of Western Europe against the Soviet Union. The Berlin Crisis of 1948-49 exacerbated these concerns and in 1949 the allies established the North Atlantic Treaty Organization (NATO), but little else was done to address the Soviet threat.

The June 1950 surprise attack on South Korea by Communist North Korea served as a catalyst and in 1951 NATO established Allied Command Europe and the Supreme Headquarters Allied Powers Europe (SHAPE). General Dwight D. Eisenhower was recalled from retirement to become the first Supreme Allied Commander Europe (SACEUR).

Even as the war in Korea raged, the U.S. reinforcements to Europe to deter the Soviet Union from similar aggression there and between 1950 and 1953, U.S. military personnel in Europe grew from 120,000 to over 400,000. To provide for national command within NATO, and to help control this build-up of forces, Eisenhower proposed a separate command for all U.S. forces in Europe. Because the senior U.S. commander would continue as Supreme Allied Commander Europe, Eisenhower recommended giving "a maximum of delegated authority" to a four-star deputy.

The first U.S. Commander-in-Chief Europe (USCINCEUR) was General Matthew B. Ridgway, former commander of Eighth Army and the Far East Command during the Korean War. USEUCOM used the Military Assistance Program to help its NATO partners build their military capabilities. USEUCOM also conducted out-of-sector operations such as a major contingency operation to Lebanon in 1958. In 1961 Berlin once again became a flashpoint when the Soviets erected a wall to stop people fleeing Communist rule.

In the early 1960s, policy disagreements emerged within NATO, and in 1966 France demanded the removal of all U.S. and NATO headquarters and forces from France. The following year SHAPE moved to Mons, Belgium, while Headquarters USEUCOM moved to Patch Barracks. USEUCOM continued to prepare for the defense of Europe and began a series of annual REFORGER (Return of Forces to Europe) exercises in 1967. Cold War crises continued, including the 1968 Warsaw Pact invasion of Czechoslovakia. But the readiness of U.S. forces in Europe slowly declined due to the Vietnam War and balance of payment problems. Troop strength in Europe fell to 265,000 by 1970.

During the 1970s, the Cold War transitioned to an era of détente and negotiations although tensions remained high as both sides modernized their conventional and nuclear forces. In the late 1970s the Soviet Union deployed SS-20 intermediate-range ballistic missiles into Eastern Europe and in 1979 invaded Afghanistan. NATO responded with a "two-track" decision to step up negotiations while deploying U.S. intermediate-range Pershing II missiles and ground-launched cruise missiles to counter the Soviet threat.

During the 1980s, the armed forces began to recover from the Vietnam War and U.S. forces in Europe grew to over 350,000. The UCP was changed in 1983 to transfer responsibility for the Middle East from USEUCOM to a new combatant command, USCENTCOM, but USEUCOM retained responsibility for the "confrontation states" of Israel, Lebanon and Syria. At the same time USEUCOM was formally assigned responsibility of Africa, south of the Sahara.

In 1989, the Soviet Union and its empire in Eastern Europe collapsed and the Cold War came to an end. In 1991 USEUCOM and its components provided forces to USCENTCOM for another out-of-sector operation, Desert Storm. USEUCOM reached out to the emerging democracies through programs such as the Joint Contact Team Program, NATO Partnership for Peace and the National Guard Bureau State Partnership Program. It was also active in peace and stability operations in the Balkans, including Bosnia, Macedonia and Kosovo. But it had to conduct these new missions with fewer assigned forces as its strength fell below 120,000.

After the September 11, 2001, terrorist attacks, USEUCOM provided major forces for operations in Afghanistan and Iraq and stepped up its efforts to protect U.S. interests in Europe and Africa.

Subsequent terrorist attacks in the USEUCOM theater in Casablanca, Madrid, London and Algiers made it clear that terrorism demanded a collective response. USEUCOM worked to build partner capacity in Europe and Africa for peacekeeping operations and deployments to Iraq and Afghanistan. USEUCOM launched Operation Enduring Freedom and Trans-Sahara in 2007 while continuing to provide rotational forces to Afghanistan and Iraq.

Subcomponents[99]

U.S. Army Europe (USAEUR)

Located in Heidelberg, Germany, USAEUR trains, equips, deploys and provides command and control of forward-deployed land forces, able to support and conduct the full spectrum of joint, and combined multi-national operations, and engagement activities. The U.S. Seventh Army and its subordinate U.S. Fifth Corps forms the basis of this subcomponent command.

U.S. Naval Forces Europe (NAVEUR)

Located in Naples, Italy, NAVEUR, the U.S. Navy component command, conducts a full range of maritime operations as well as theater security cooperation operations with NATO allies. The U.S. Sixth Fleet, which is also a subcomponent of AFRICOM, forms the basis of this subcomponent command.

U.S. Air Forces in Europe (USAFE)

Located at Ramstein Air Force Base in Germany, USAFE, the Air Force component command, operates from five main operating bases that supports nine air wings. These wings provide a full spectrum of air support, including strategic airlift as well as air support to ground forces and intelligence, surveillance, and reconnaissance support. The Third Air Force forms the basis of this subcomponent command.

U.S. Marine Corps Forces Europe (MARFOREUR)

Located in Stuttgart, Germany, MARFOREUR employs pre-positioned assets to rapidly deploy expeditionary forces and equipment and conduct a wide array of operations, including building partner capacity.

Special Operations Command Europe (SOCEUR)

Located at Patch Barracks, Stuttgart, Germany, SOCEUR provides flexibility throughout a full range of military operations including combat, special operations, humanitarian assistance, noncombatant evacuations and joint-combined military operations.

Ongoing Operations[100]

Support to NATO's International Security Assistance Force (ISAF) in Afghanistan[101]

USEUCOM's support to ISAF is significant. About 80% of the non-U.S. countries deployed to Afghanistan are from the European theater. USEUCOM's support to ISAF is primarily focused on preparing these nation's forces for deployment to Afghanistan. These supporting activities include sending mobile planning teams to assess partner nation equipment and training requirements and working with these countries to develop a comprehensive pre-deployment plan.

Multi-National Joint and Interagency Exercises

Multi-national and joint interagency exercises constitute the most significant form of peacetime interaction with NATO allies and other partners. In 2010, USEUCOM conducted 33 major exercises involving almost 50,000 U.S., allied, and partner nation personnel from 40 nations. These exercises focused on preparing partner nations for ongoing coalition operations including ISAF, enhancing overall NATO interoperability, and improving NATO's military interoperability with Israel.

Exercises in the Baltics, Balkans, and Caucasus

In support of NATO initiatives, USEUCOM provides U.S. forces for nine NATO and Partnership for Peace training events in the Baltics. In addition, USEUCOM provides forces for two major exercises in the Balkans which not designed to help to improve these nation's military capabilities but also to ease regional tensions.

Selected Current Issues[102]

Afghanistan

USEUCOM currently has about 12,000 U.S. troops forward deployed to Afghanistan serving in a variety of capacities. As previously noted, USEUCOM plays a central role in training and deploying non-U.S. forces and the USEUCOM Commander, in his SACEUR role, is a central figure in persuading NATO allies to provide troops and resources for ISAF. As overall U.S. involvement in Afghanistan begins to wind down, NATO involvement will also likely follow suit and, in some instances, some NATO nations might opt for an abrupt end to their support to ISAF. USEUCOM's and SACEUR's primary challenge will likely be to develop plans for and manage a smooth transition as NATO forces draw down and transition security functions to the Afghan National Army. It is possible USEUCOM and NATO might have some residual support requirements in Afghanistan after the transition and these efforts will also require resources and management.

The Balkans

USEUCOM has participated in NATO operations in the Balkans since the 1995 Dayton Peace Accords. At the height of Operation Joint Endeavor in 1996, the United States had over 20,000 troops in the Balkans.

While current USEUCOM troop commitments to the Balkans are negligible, continued engagement in the region is viewed by most as essential in keeping the region stable. As such, USEUCOM participation in NATO exercises and training teams in this region is deemed essential not only to improve capabilities and diffuse tensions but also to signify U.S. long-term interest and commitment to the Balkan region.

Missile Defense[103]

USEUCOM notes an increasing and expanding ballistic missile threat to USEUCOM's area of focus, citing missile-related activities in Iran and Syria, as well as those of nonstate actors such as Hizbollah. In response to this threat, the U.S. and Poland have initiated a cooperative air and missile defense partnership whereby the U.S. rotates Patriot anti-missile and anti-aircraft batteries to Poland on a quarterly basis and conducts training with their Polish counterparts. In September 2011, Romania agreed to the stationing of 24 interceptor missiles on Romanian soil and Turkey agreed to accept a sophisticated U.S. radar, possibly by the end of the year.[104]

These events as well as plans to expand coverage to other countries, has elicited opposition from Russia and has complicated NATO's relationship with Moscow. USEUCOM's challenge will likely be to continue to play a role in developing European ground-based missile defense while at the same time maintaining an effective military relationship with Russia under potentially trying circumstances.

U.S. Northern Command (USNORTHCOM)
Website: http://www.northcom.mil

Mission[105]

"USNORTHCOM's mission is to conduct homeland defense, civil support and security cooperation to defend and secure the United States and its interests. USNORTHCOM was established Oct. 1, 2002, to provide command and control of DOD homeland defense efforts and to coordinate defense support of civil authorities.

USNORTHCOM's area of operation includes air, land and sea approaches and encompasses the continental United States, Alaska, Canada, Mexico and the surrounding water out to approximately 500 nautical miles.

It also includes the Gulf of Mexico, the Straits of Florida, portions of the Caribbean region to include The Bahamas, Puerto Rico, and the U.S. Virgin Islands. The commander of USNORTHCOM is responsible for theater security cooperation with Canada, Mexico, and The Bahamas."

North American Aerospace Defense Command (NORAD)[106]

"The commander of USNORTHCOM also commands NORAD, a joint U.S.-Canadian command responsible for aerospace warning, aerospace control, and maritime warning for Canada, Alaska and the continental United States. For the aerospace warning mission, the commander of NORAD provides an integrated tactical warning and attack assessment to the governments of Canada and the United States. To accomplish the aerospace control mission, NORAD uses a network of satellites, ground-based radar, airborne radar and fighters to detect, intercept and, if necessary, engage any air-breathing threat to Canada and the United States. In conjunction with its aerospace control mission, NORAD assists in the detection and monitoring of aircraft suspected of illegal drug trafficking. This information is passed to civilian law enforcement agencies to help combat the flow of illegal drugs into North America. The Command has developed an initial concept for implementing the new maritime warning mission."

Missile Defense

USNORTHCOM is the combatant command responsible for the operation of the Ground-Based Midcourse Defense System (GMD) designed to defend the United States against the threat of a limited ballistic missile attack from nations such as North Korea and Iran.

Unique Civil Support Mission[107]

USNORTHCOM's civil support mission includes domestic disaster relief operations during fires, hurricanes, floods and earthquakes. Support also includes counter-drug operations and managing the consequences of a terrorist event employing a weapon of mass destruction. The command provides assistance to a nonmilitary Primary Agency when tasked to do so by DOD. Per the Posse Comitatus Act,[108] military forces can provide civil support, but cannot become directly involved in law enforcement.[109]

History[110]

On April 17, 2002, DOD announced the establishment of USNORTHCOM to consolidate under a single unified command those existing homeland defense and civil support missions that were previously executed by other military organizations. On May 8, 2002, U.S. Air Force General Ralph E. Eberhart, the commander of the North American Aerospace Defense Command (NORAD) and U.S. Space Command, was nominated to be the first commander of USNORTHCOM. USNORTHCOM attained initial operational capability on October 1, 2002.

USNORTHCOM provided support in response to the Space Shuttle Columbia disaster in February 2002, and in May 2003 participated in a comprehensive terrorism response exercise States—Top Officials 2, or TOPOFF 2. USNORTHCOM conducted Exercise Determined Promise 03 in Clark County, NV, and Colorado Springs, CO. This major exercise was designed to evaluate the command's ability to command and control multiple homeland defense and defense support of civil authorities missions simultaneously. Following Exercise Determined Promise 03, USNORTHCOM announced full operational capability on September 11, 2003.

In February 2004, USNORTHCOM conducted Exercise Unified Defense 04. This major exercise allowed the USNORTHCOM, Fifth Army, Joint Task Force Alaska and associated units to practice the homeland defense and defense support to civil authorities missions. The exercise involved the Department of Homeland Security and more than 50 federal, state and local organizations. During the summer of 2004, USNORTHCOM supported interagency efforts to deter and defeat any possible threats against several National Security Special Events. Exercise Determined Promise 04, conducted in August 2004, tested USNORTHCOM's ability to assist civil and federal authorities in a coordinated response to simulated chemical, radiological, and explosive hazards, conducted in California and Virginia. In the same month, USNORTHCOM supported the Federal Emergency Management Agency's efforts to provide relief to areas in Florida most impacted by Hurricane Charley.

In September 2005, as directed by the secretary of defense and in accordance with the National Response Plan, USNORTHCOM supported the Department of Homeland Security and Federal Emergency Management Agency (FEMA) and other federal agencies in disaster relief efforts in the aftermath of Hurricane Katrina. More than 21,400 active-duty servicemembers and 45,700 Army and Air National Guard members supported the effort in the U.S. Gulf Coast.

In May 2006 USNORTHCOM participated in Exercise Ardent Sentry 06, which involved numerous federal, provincial, state and local agencies in Canada and the United States. The exercise required participants to respond to simulated terrorist activities and manage the consequences of a range of simulated man-made and natural disasters. Exercise Ardent Sentry 06 helped military and civilian officials prepare to respond to a variety of national crises.

On July 4, 2006, the Democratic Peoples Republic of Korea launched six ballistic missiles, including a long-range Taepodong-2 missile. USNORTHCOM personnel were immediately able to detect the launch of all the missiles. While Ground-based Midcourse Defense System interceptors at Fort Greely, AK, and Vandenberg Air Force Base, CA, were operational during the launches, top officials from the command were able to quickly determine the missiles posed no threat to the United States or its territories.

Subcomponents[111]

Joint Force Headquarters National Capital Region (JFHQ-NCR)

"JFHQ-NCR, based at Fort McNair, Washington, D.C. is responsible for land-based homeland defense, defense support of civil authorities (DSCA), and incident management in the National Capital Region. JFHQ-NCR is responsible for protecting the District of Columbia and neighboring counties and cities of Maryland and northern Virginia. JFHQ-

NCR draws together the existing resources of the Army, Navy, Air Force, Marine Corps, Coast Guard and NORAD into a single point headquarters for planning, coordination and execution of the mission in the National Capital Region."

Joint Task Force Alaska (JTF-AK)

"JTF-AK is headquartered at Elmendorf Air Force Base, Alaska. JTF-AK's mission is to, in coordination with other government agencies, deter, detect, prevent and defeat threats within the Alaska Joint Operations Area (AK JOA) in order to protect U.S. territory, citizens, and interests, and as directed, conduct Civil Support operations."

Joint Task Force Civil Support (JTF-CS)

"JTF-CS, was originally formed as a standing joint task force under USJFCOM. JTF-CS was transferred to USNORTHCOM when USNORTHCOM was established October 1, 2002. The task force consists of active, Guard and Reserve military members drawn from all service branches, as well as civilian personnel, who are commanded by a federalized (Title X) National Guard general officer.

JTF-CS plans and integrates DOD support to the designated Primary Agency for domestic chemical, biological, radiological, nuclear, or high-yield explosive (CBRNE) consequence management operations. When approved by the secretary of defense and directed by the commander of USNORTHCOM, JTF-CS deploys to the incident site and executes timely and effective command and control of designated DOD forces, providing support to civil authorities to save lives, prevent injury and provide temporary critical life support."

Joint Task Force North (JTF North)

"JTF North, based at Biggs Army Airfield, Fort Bliss, TX, is the DOD organization tasked to support our nation's federal law enforcement agencies in the interdiction of suspected transnational threats within and along the approaches to the continental United States.

Transnational threats are those activities conducted by individuals or groups that involve international terrorism, narcotics trafficking, alien smuggling, weapons of mass destruction, and the delivery systems for such weapons that threaten the national security of the United States."

Army North (ARNORTH)

"ARNORTH is the Army component of NORTHCOM and is located at Fort Sam Houston, TX. ARNORTH's mission is to conduct homeland defense, civil support operations and theater security cooperation activities. ARNORTH is responsible for developing and unifying the military response capability for chemical, biological, radiological, nuclear and high-yield explosives (CBRNE) incidents. In addition, the Civil Support Readiness Directorate trains National Guard Weapons of Mass Destruction Civil Support Teams, which are state first responders for chemical, biological, radiological, nuclear or high-yield explosive incidents."

Air Force North (AFNORTH)

"Headquartered at Tyndall Air Force Base, near Panama City, FL, 1st Air Force is assigned to Air Combat Command. It has the responsibility of ensuring the air sovereignty and air defense of the continental United States. As the CONUS geographical component of the bi-national North American Aerospace Defense Command, it provides airspace surveillance and control and directs all air sovereignty activities for the continental United States Fleet Forces Command (USFF)."

U.S. Fleet Forces Command (USFF)

"USFF is the Navy component of USNORTHCOM and is located at Norfolk, VA. USFF's mission is to provide maritime forces prepared to conduct homeland defense, civil support operations and theater security cooperation activities when directed by USNORTHCOM. Additionally, USFF has responsibilities to generate ready Navy forces for assignment to global Regional Combatant Commanders, execute the Fleet Response Plan (FRP) using the Fleet Training Continuum, articulate to the Chief of Naval Operations the integrated Fleet warfighting requirements as coordinated with all Navy Component Commanders, and provide operational planning support to USTRATCOM."

Ongoing Operations

USNORTHCOM's and NORAD's missions of homeland defense, air and missile defense, and maritime warning involve a multitude of continuous operations in a variety of domains. These operations are best described as monitoring, detection, and warning, and, in the case of air-breathing threats, interception. According to the USNORTHCOM commander:

> Our daily efforts include countering terrorism and transnational criminal organizations, preparing to support our federal and state partners in the wake of a natural or manmade disaster, air defense against both internal and external threats, maritime and ballistic missile defense and of course, a growing focus on the Arctic.[112]

Selected Current Issues[113]

Countering Transnational Criminal Organizations[114]

Mexican transnational criminal organizations (TCOs) and their involvement in the illicit trafficking of drugs, weapons, money, and human beings into the United States is one of USNORTHCOM's primary homeland security concerns. USNORTHCOM, in keeping with U.S. law and working with and through other U.S. government agencies, is working with the Mexican government to defeat the TCOs. USNORTHCOM efforts include providing the Mexican military with material solutions as well as sharing operational insights and experiences. In addition to efforts along the southern U.S. border, USNORTHCOM has been providing support to U.S. law enforcement agencies along both southern and northern borders. This support includes sensors, radar, forward-looking infrared, and manned and unmanned aerial border surveillance. Because access to these platforms is not unlimited, concerns exist that if a greater level of support is required along one border, assets available for the other border might become constrained.

The Arctic[115]

Because of the growing geo-strategic importance of the Arctic, the USNORTHCOM Commander has designated the Arctic as a key focus area. Along these lines, USNORTHCOM is currently examining how to support other U.S. government agencies in the region with search and rescue assets, humanitarian assistance, disaster response, and law enforcement. As part of this examination, USNORTHCOM has identified deficiencies in all-domain awareness, communications, infrastructure (including a deepwater port), mobility (including an adequate national icebreaking capability), search and rescue enabling capabilities, Arctic Ocean charting, and the ability to observe and forecast Arctic environmental change.

U.S. Pacific Command (USPACOM)
Website: http://www.pacom.mil

Mission[116]

"USPACOM protects and defends, in concert with other U.S. government agencies, the territory of the United States, its people, and its interests. With allies and partners, USPACOM is committed to enhancing stability in the Asia-Pacific region by promoting security cooperation, encouraging peaceful development, responding to contingencies, deterring aggression, and, when necessary, fighting to win. This approach is based on partnership, presence, and military readiness."

USPACOM's AOR covers half of the earth and is home to three billion people living in three dozen countries with five of these nations being U.S. allies and with many more important economic and security partners. USPACOM's AOR contains the world's three largest economies and almost one-third of U.S. two-way trade in goods and services. In addition, much of the world's trade and energy that fuels the global economy transits Asia's sea and air lines of communication.

History[117]

USPACOM was established as a unified command on January 1, 1947, and is the oldest and largest of the United States' COCOMs. The present USPACOM includes areas originally assigned to two other unified commanders. The Far East Command, which had been established on January 1, 1947, was disestablished on July 1, 1957, and all its responsibilities were assumed by the Pacific Command. That same day the command assumed some of the responsibilities of the Alaskan Command and individual Army and Air Force component commands for the Pacific also were established in Hawaii.

Added responsibilities were assigned to USPACOM on January 1, 1972, for military forces and elements in the Indian Ocean, Southern Asia, and the Arctic. The Pacific Command's AOR was further expanded on May 1, 1976, to the east coast of Africa. This enlarged the Pacific Command to more than 50% of the earth's surface, an area of over 100 million square miles.

Another enlargement of the USPACOM area took place in October 1983, when CINCPAC was assigned responsibility for the People's Republic of China, the Democratic People's Republic of Korea, Mongolia, and the Republic of Madagascar. CINCPAC was also redesignated Commander in Chief, U.S. Pacific Command (USCINCPAC).

A new Alaskan Command (ALCOM) was established on July 7, 1989, at Elmendorf Air Force Base, Alaska, as a subordinate unified command responsible to USCINCPAC. This placed the defense of Alaska and its surrounding waters under the leadership of one commander, providing a unity of command absent from the state since the early 1970s.

From 1989 through 2000, three UCPs slightly reduced USPACOM's AOR. With the focus of attention shifting to the Middle East, the August 16, 1989 plan assigned responsibility for the Gulf of Oman and Gulf of Aden to Commander, USCENTCOM. The January 1, 1996, plan transferred the Seychelles and adjacent waters to USCENTCOM. On October 1, 2000, responsibility for Indian Ocean waters off Tanzania, Mozambique, and South Africa was transferred from USPACOM to USEUCOM.

The UCP changed as a result of the events of September 11, 2001, and the ensuing war on terrorism, as well as the new defense strategy articulated in the 2001 Quadrennial Defense Review. For the first time the entire surface of the earth was divided among the various unified commands. A new USNORTHCOM was created for homeland security and other changes in the various commands' responsibilities resulted in significant changes for USPACOM. The West Coast of North America was reassigned from USPACOM to USNORTHCOM. While Alaska was included in the reassignment to USNORTHCOM, Alaskan Command forces remained assigned to USPACOM in the Forces for Unified Commands Memorandum. Antarctica was also added to USPACOM's AOR. Approved in April 2002, the new UCP became effective October 1, 2002.

The 2008 UCP, signed on December 17, 2008, documented the transfer of all areas of the Indian Ocean previously assigned to USPACOM west of 68 degrees east to the newly established USAFRICOM. As a result, four island countries off the east coast of Africa that were formerly assigned to PACOM were reassigned to AFRICOM: Comoros, Madagascar, Mauritius, and Reunion."

Subcomponents[118]

U.S. Army Pacific (USARPAC)

Located in Fort Shafter, Hawaii, USARPAC is the Army's component command in the Pacific and supplies Army forces for full-spectrum security operations. USARPAC is the most forward deployed unit in the Army still on U.S. soil in Hawaii.

U.S. Forces Korea (USFK)

USFK is a subcommand within USPACOM responsible to U.S. forces in Korea. While this is a joint headquarters, command has traditionally been held by a three-star, U.S. Army general.

U.S. Eighth Army

The U.S. Eighth Army operates in conjunction with USFK and the United Nations Command in Korea. U.S. Eighth Army's stated mission is described as

> Eighth Army supports deterrence of North Korea aggression against the Republic of Korea. Should deterrence fail, Eighth Army supports Non-combatant Evacuation Operations (NEO), transitions to hostilities, generates combat power to support Commander in Chief United Nations Command/USFK's campaign, and provides combat support and combat service support to assigned, attached, and other designated forces within the Korea Theater of Operation and on order, conducts combat operations.[119]

U.S. Pacific Fleet (PACFLT)

PACFLT consists of the California-based Third Fleet and the Fifth Fleet in Japan. It is the world's largest fleet command responsible for 100 million square miles, more than half the Earth's surface, from the West Coast of the United States into the Indian Ocean. PACFLT consists of approximately 180 ships, nearly 2,000 aircraft and 125,000 Sailors, Marines and government civilian employees.[120]

U.S. Pacific Air Force (PACAF)

PACAF is headquartered at Hickam Air Force Base, Hawaii, where it plans, conducts, and coordinates defensive and offensive air operations in the Asian and Pacific region. PACAF's components consists of the Seventh Air Force in South Korea, the Fifth Air Force in Japan, the Eleventh Air Force in Alaska, and the Thirteenth Air Force in Guam.

U.S. Marine Forces Pacific (MARFORPAC)

MARFORPAC, the Marine component headquarters, includes the First Marine Expeditionary Force in California and the Third Marine Expeditionary Force based in Okinawa.

Special Operations Command Pacific (SOCPAC)

SOCPAC, located at Camp H. M. Smith, Oahu, Hawaii, is a sub-unified command and serves as the SOF component command for USPACOM.

Special Operations Command Korea (SOCKOR)

SOCKOR, located at Camp Kim in Yongsan, Korea, is the Theater Special Operations Command responsible for special operations on the Korean peninsula and, when established, the Korean Theater of Operations.

Other Major USPACOM Organizations[121] Joint Intelligence Operations Center (JIOC)

The JIOC is the central clearing house for intelligence throughout the theater and is responsible for managing intelligence requirements at the strategic level and supports USPACOM Subcomponents and Subordinate Commands.

Asia-Pacific Center for Security Studies (APCSS)

APCSS supports USPACOM security cooperation and capacity-building efforts by means of international executive education and specialized assistance programs that are intended to both educate and foster relationships between key regional security officials.

Joint POW/MIA Accounting Command (JPAC)

JPAC's mission is to achieve the fullest possible accounting of all Americans missing as a result of past conflicts.

Joint Interagency Task Force West (JIATF-West)

JIATF-West is the USPACOM executive agent for countering drug-related transnational crimes in the Asia-Pacific region primarily by supporting U.S. law enforcement agencies operating in the region.

Ongoing Operations

While USPACOM has significant on-the-ground presence in Korea as well as a variety of naval and air activities throughout its AOR, its primary focus is exercise and engagement programs.

USPACOM's current program consists of 18 major exercises involving joint military forces as well as other U.S. government agencies. These exercises are conducted with 27 of 36 USPACOM partner nations.

On the operational side, USPACOM played a crucial role in helping Japan in the aftermath of the March 11, 2011, earthquake and resultant tsunami which devastated parts of Japan.

USPACOM and its subordinate commands provided direct disaster relief support on the ground, at sea, and in the air.

Selected Current Issues

In his April 2011 testimony to the Senate Armed Services Committee, the USPACOM commander indicated current major issues in his AOR.[122] These issues included the threat to the United States and its allies posed by North Korea's nuclear and missile capabilities, its proliferation of weapons of mass destruction and associated technologies, and its potential for instability. [123]

Another issue was transnational violent extremist organizations (VEOs) undermining stability and threatening Allies and emerging partners.

The USPACOM commander voiced concern with China's significant military modernization associated with its unclear intent.[124]

He also noted that territorial disputes, and increasingly assertive actions to resolve them, present the potential for conflict and instability.

Cyber threats and transnational criminal activity—to include piracy and trafficking in narcotics and persons—were cited as growing areas of concern.

Finally, humanitarian crises such as pandemics and famines, as well as natural disasters such as tsunamis, earthquakes, and volcanoes; and environmental degradation presented unique challenges to USPACOM.

U.S. Southern Command (USSOUTHCOM) Website: http://www.southcom.mil/appssc/index.php

Mission[125]

"USSOUTHCOM is responsible for providing contingency planning, operations, and security cooperation for Central and South America, the Caribbean (except U.S. commonwealths, territories, and possessions), Cuba; as well as for the force protection of U.S. military resources at these locations. USSOUTHCOM is also responsible for ensuring the defense of the Panama Canal and canal area."

History[126]

During World War II, the Roosevelt administration established the U.S. Caribbean Defense Command (1941-1947), a prototype unified military organization, to defend the Panama Canal and surrounding area. The command organized and implemented an active system of regional defense, including antisubmarine and counterespionage operations.

Located in Panama, the U.S. Caribbean Defense Command also established military training missions in Latin America; distributed military equipment to regional partners through the Lend Lease program; and opened U.S. service schools to Latin American soldiers, sailors, and airmen. At the height of the war, U.S. military planners assigned 130,000 uniformed personnel to duty stations in Latin America and the Caribbean. Roughly half of those forces were under the direct control of the U.S. Caribbean Defense Command.

In 1947, U.S. strategists adopted a national security plan that transformed the wartime headquarters into the U.S. Caribbean Command. Beyond defending the Panama Canal, it assumed broad responsibilities for inter-American security cooperation in Central and South America. During the 1950s, defense officials also removed the Caribbean basin from the U.S. Caribbean Command's area of focus.

In the event of a global war with the communist powers, they reasoned, U.S. Atlantic Command, based in Norfolk, VA, needed the Caribbean basin to conduct hemispheric antisubmarine operations.

By 1960, the U.S. Caribbean Command carried a name that incorrectly described its geographic interests, Central and South America. The Kennedy administration changed the name to USSOUTHCOM on June 6, 1963. During the 1960s, the USSOUTHCOM mission involved defending the Panama Canal, contingency planning for Cold War activities, and the administration of the U.S. foreign military assistance program in Central and South America. In particular, USSOUTHCOM personnel undertook civic-action projects with partner nation forces to accelerate regional development. During the 1970s the Joint Chiefs of Staff recommended disestablishing the command to trim the U.S. military presence abroad.

For political reasons, the command narrowly survived, albeit with limited responsibilities and resources. In the 1980s, internal conflicts in El Salvador, Nicaragua, and elsewhere rekindled U.S. military interest in Latin America and the Reagan administration revitalized USSOUTHCOM.

When the Cold War ended, the command, like other U.S. military organizations, entered a period of dramatic change.

In rapid succession, USSOUTHCOM was assigned responsibility for counter-drug operations, expanded its area of geographic focus to include the Caribbean, and enhanced its capacity for humanitarian missions. In September 1997, USSOUTHCOM moved to Miami, FL.

Subcomponents[127]

U.S. Army South (ARSOUTH)

ARSOUTH is located at Ft. Sam Houston, Texas, where its primary mission is to support regional disaster and counterdrug operations. ARSOUTH also is responsible for oversight, planning, and logistical support for humanitarian and civic assistance projects throughout the region.

U.S. Naval Forces Southern Command/U.S. Fourth Fleet (COMUSNAVSO /COMFOURTHFLT)

COMUSNAVSO/COMFOURTHFLT is located in Mayport Naval Base in Florida and supports USSOUTHCOM with a full range of naval capabilities. Its primary responsibility is to provide sea-based forward presence to ensure freedom of maneuver as well as developing cooperative relationships with partners in the region.

Air Forces Southern/ Twelfth Air Force (AFSOUTH)

AFSOUTH is located at Davis-Monthan Air Force Base in Arizona and is responsible for Air Force forces in the region. AFSOUTH serves as the executive agent for forward operating locations in the region and provides joint and combined radar surveillance and intra-theater airlift.

U.S. Marine Forces South (USMARFORSOUTH)

USMARFORSOUTH is located in Miami, Florida, and advises USSOUTHCOM on the proper employment and support of Marine forces operating in the region. In addition, USMARFORSOUTH conducts deployment/redeployment planning and supervises mission execution for assigned Marine forces.

Special Operations Command South (USSOCSOUTH)

USSOCSOUTH is located near Miami, Florida, and provides primary theater contingency response forces and plans for and conducts special operations in support of USSOUTHCOM. USSOCSOUTH can also serve as a Joint Special Operations Task Force when required.

USSOUTHCOM Task Forces and Direct Reporting Units[128]

Joint Task Force Bravo (JTF-Bravo)

JTF-Bravo is located at Soto Cano Air Base, Honduras, and operates a forward, all weather, day/night C-5 Galaxy-capable air base. JTF-Bravo organizes multilateral exercises and, with partner nations, supports humanitarian and civic assistance, counterdrug, contingency and disaster relief in Central America.

Joint Task Force Guantanamo (JTF–Guantanamo)

JTF-Guantanamo is located at the U.S. Naval Station Guantanamo, Cuba, and conducts detention and interrogation operations in support of world-wide U.S. counterterrorism operations.

Joint Interagency Task Force South (JIATF South)

JIATF-South is located in Key West, Florida and is an interagency task force that integrates and synchronizes U.S, counterdrug operations and is responsible for the detection and monitoring of suspect air and maritime drug activity in the region. JIATF South works in coordination with USNORTHCOM's JTF North on a variety of counterdrug and counter trafficking operations.

Center for Hemispheric Defense Studies (CHDS)

CHDS is located in Washington, DC, and provides education, outreach, and research and knowledge-sharing activities on defense and policy making with regional military and political leaders.

Ongoing Operations[129]

Exercises and Military-to-Military Activities

USSOUTHCOM is involved in a variety of exercises and military-to-military operations in support of the Theater Engagement Plan. On an annual basis, USSOUTCOM conducts medical readiness training exercises, engineering exercises, and disaster relief and humanitarian assistance exercises. In 2010, in response to the January 12, 2010, earthquake in Haiti, USSOUTCOM was placed in charged of Operation Unified Response to support the U.S. government's response to the disaster.

While the operation concluded in May 2010, USSOUTHCOM continues an annual exercise, New Horizon Haiti, that provides targeted humanitarian and civic assistance. In addition USSOUTHCOM conducts seven annual military exercises designed to facilitate interoperability, build capabilities, and provide opportunities to share best practices with regional military and security forces.

Task Force-Oriented Activities

JTF-Guantanamo continues to serve as a detention and interrogation center for suspected terrorists. JTF-Bravo and JIATF-South are involved in a wide variety of day-to-day activities and operations designed to counter illicit trafficking of people, narcotics, money, and weapons. In addition to operations against

Transnational Criminal Organizations, USSOUTHCOM task forces also work to counter violent extremist organizations from the Middle East which have been active in Latin America and the Caribbean and are considered a potential threat.

Selected Current Issues[130]

Counter-Trafficking

The USSOUTHCOM Commander noted that illicit trafficking of drugs, weapons, and people and their associated TCOs constitute the primary threat to regional security. Working in conjunction with regional partners, USSOUTCOM is combating these criminal organizations through demand reduction; eradication and regulation of source materials; suppression of money laundering; and interdiction of illegal shipments as they transit to the U.S. and other end-user countries. These efforts not only involve regional partners but also various U.S. Interagency offices.

Natural Disasters, Poverty, and Violence

The USSOUTHCOM Commander testified that natural disasters, poverty, and violence in the region have a negative impact on regional security and stability.

Widespread and frequent natural disasters in the region have worsened economic and social conditions in countries that can ill-afford these types of setbacks and when governments can not make discernable progress recovering in the aftermath of these events, citizens loose faith in government.

While economic conditions in some countries have improved, poverty, particularly in Central America, creates conditions for social stagnation.

These social conditions create openings for criminal organizations to recruit new members who both undermine legitimate governance and contribute to increasing violence against private citizens.

Extra-Regional Actors

While the USSOUTHCOM Commander noted there are economic benefits for countries in his AOR in establishing or renewing relationships with extra-regional actors such as China, Russia, and Iran, it also presents a number of challenges. China, for example, has sold light attack and training aircraft to Venezuela and Bolivia and air surveillance radars to Venezuela and Ecuador.

Russian arms transfers with Latin American in 2009 were valued at $5 billion and Russian sales of automatic weapons and man-portable air defense systems to Venezuela are viewed as troubling.

Iran has nearly doubled the number of its embassies in the region over the past decade and hosted three regional heads of state in 2010.

In addition to extra-regional state actors, violent extremist organizations from the Middle East are active in Latin America and the Caribbean and are involved in fund-raising activities to finance worldwide activities.

POTENTIAL ISSUES FOR CONGRESS

Congress is presented with a wide range of national security policy issues that are impacted by the provisions of the UCP as well as the COCOM construct. As the U.S. arguably moves to a post- Iraq/Afghanistan era where global military operations against terrorists could be the new "steady state," it might prove prudent to re-examine the UCP and COCOMs. As new threats such as cyber attacks and TCOs take center stage and new international actors such as China and India emerge while established actors such as Russia and Iran transition to different types of security challenges, such a re-examination could serve to increase the efficacy of U.S. national security policy.

Is Greater Interagency Involvement in the UCP Process Needed?[131]

In the Fall 2010 edition of the *Interagency Journal,* former U.S. Ambassador Edward Marks noted

> The geographic commands have essentially two tasks: war planning and fighting, and military engagement programs. Both tasks remain, and will always remain, fundamental responsibilities of the Department of Defense and the military services. However while the war planning and fighting responsibility obviously remains uniquely a duty of the Department of Defense and the military services, the engagement programs no longer can be handled as a discrete military activity. In today's world, military engagement programs with other countries can only be seen as part of the overall engagement activity of the U. S. government. The so-called "nexus" of security challenges—terrorism, narcotics, smuggling, international criminal networks, etc.—can no longer be managed as single agency programs but must be integrated into "whole of government" programs.[132]

The concept of a "whole of government approach to national security" has taken on renewed emphasis since September 11, 2001. Past and current senior military leadership have repeatedly called for greater participation in national security matters from other U.S. government agencies, even going so far as to publically advocate for greater funding levels for the State Department and U.S. Agency for International Development (USAID) so they can play a greater role in military operations.

In the current strategic environment, COCOMs are being faced with security challenges that fall outside the traditional military realm. One such challenge, transnational criminal organizations or TCOs, is a stated concern of Combatant Commanders both in a domestic and international context. In this regard, if TCOs are expected to become a central security issue for COCOMs and the President and DOD include TCO-related responsibilities in the UCP, enhanced interagency involvement in the UCP process from the Justice Department, other U.S. law enforcement entities and others could prove to be beneficial.

It can be argued while greater resources for other U.S. government agencies are important, of equal importance is participation in the UCP process. It has been noted that military engagement programs are at the forefront of geographical COCOM's responsibilities and as hostilities in Iraq and Afghanistan diminish over time, and the U.S. defense budget decreases, military engagement could become the primary focus of all geographical COCOMs.

Under the current UCP development process, the U.S. Interagency has a degree of visibility but participation is limited. While Interagency participation in developing regional war plans might not be appropriate, a greater role in planning for military engagement activities might not only enhance these programs but might also identify areas of redundancy with other U.S. government regional engagement programs. This enhanced role could include more Interagency representatives in the early stages of the UCP review and development process and increasing military presence in key Interagency positions, particularly directorates that are responsible for strategic planning and resourcing.

While the Interagency might welcome the opportunity to play a greater role, DOD might be less than enthusiastic with including a greater role for other U.S. government agencies in what it likely considers fundamental strategic military planning.

In this regard, Congress might consider an in depth examination of the UCP development process.

This examination could focus on the current level of Interagency participation and identifying areas in the process where greater Interagency involvement could be beneficial.

Has U.S. Foreign Policy Become "Too Militarized" as a Result of the Geographic COCOMs?

In September 2000, Washington Post reporter Dana Priest published a series of articles whose central premise was Combatant Commanders yielded an inordinate amount of political influence within the countries in their areas of responsibility and "had evolved into the modern-day equivalent of the Roman Empire's proconsuls—well-funded, semi-autonomous, unconventional centers of U.S. foreign policy."[133]

Some national security experts consider this series as the catalyst of the continuing debate as to whether or not COCOMs have assumed too much influence overseas, thereby diminishing the roles other U.S. government entities play in foreign and national security policy.

Despite the post-September 11, 2011, ascendancy of the Interagency in foreign policy and national security matters, the debate over the COCOM's role continues. In 2007, testimony from Mark Malan from Refugees International before the Senate Subcommittee on African Affairs of the Foreign Relations Committee he noted

> In some parts of the world, like Iraq and Afghanistan, the face of U.S. foreign policy is clearly a military one. In Africa, the DOD appears to be putting a civilian mask on the face of a combatant command, with its marketing pitch for USAFRICOM. This disingenuous strategy is not working.
>
> The veneer of the mask is simply too thin, and attempts to patch the holes that have emerged—by telling us "what AFRICOM is not about" and re-emphasizing a humanitarian and developmental role for the U.S. military in Africa—simply make the face of U.S. foreign policy much shadier.
>
> The notion of a benign U.S. combatant command is an enigma to those who clearly understand (and acce pt) the need for the U.S. to secure access to Africa's natural resources, especially oil; and to establish bases from which to destroy networks linked to Al-Qaeda. When the U.S. promotes a combatant military command in terms of development and humanitarianism, Africans will inevitably suspect that the true story is being kept from them.[134]

The assertion that COCOMs have usurped other U.S. government entities in the foreign policy arena may deserve greater examination. Geographic Combatant Commanders generally agree their role is more political than military.

A former USEUCOM and Supreme Allied Commander, Europe (SACEUR) estimated he spent about 70% of his time on political-military issues, despite having ongoing combat operations in the Balkans.[135]

USCENTCOM commanders have reportedly spent a significant amount of time meeting with the senior Iraqi and Afghan political leadership over the past ten years discussing issues of building and maintaining armed forces, civil-military relations, and other national security matters.

While these discussions might not conform to what have been traditionally considered war fighting-related topics, the complexities of U.S. involvement in these two countries suggests Combatant Commanders have been required to play a more pronounced political role.

Some U.S. government officials suggest the Combatant Commander/State Department relationship, as it currently exists, has proven beneficial. A former Under Secretary of State for Political Affairs noted he was

> a huge fan of the [regional commanders]. I was the ambassador to Turkey; in EUCOM, when the deputy [commander, the commander], and I were on the same page—there was nothing we couldn't achieve. In 6 years in Turkey as [deputy chief of mission] and ambassador, there was never a single conflict.
>
> Now, I'm dealing with Colombia; I've made five of my six visits with SOUTHCOM's commander. We do everything together. Yes, someone could goof. But the system works wonderfully—the [regional commanders] are some of the finest America has to offer. When the [commander] and ambassador are on the same page, it's a very powerful combination. I'm a complete believer.[136]

Congress has examined aspects of this COCOM-State Department relationship in terms of the broader topic of civil-military relations as well as how it pertains to USAFRICOM and its role in U.S. foreign policy. In a broader context Congress might wish to consider the role Geographic COCOMs play in U.S. foreign policy abroad. This consideration could take into account more than just the State Department, but also other U.S. government agencies that play a foreign policy role.

While presence and access to resources have been cited as positive attributes for COCOM involvement overseas, it is possible a reallocation of resources might put a more "civilian" face on U.S. engagement and development efforts, possibly resulting in greater acceptance and efficacy in regions that are sensitive to U.S. military presence.

In examining the respective roles of COCOMs, the State Department, and others, it might be possible to identify both areas of redundancy as well as areas requiring greater emphasis, thereby enhancing overall U.S. effectiveness in political-military relations with nations in respective regions. With many experts predicting shrinking or flat U.S. military and State Department budgets over the next few years, such an examination might lead to a more cost effective approach to U.S. foreign policy.

Are There Other Regions or Functions That Merit a Separate COCOM?[137]

While Geographic COCOMs suggest their regional perspective is their primary virtue, others argue the "strict geographic regionalism" the COCOMs were aligned under is no longer how the world is organized.[138] These critics contend globalization at one end and localism (tribalism) at the other end has made the Geographic COCOM construct less than ideal.[139]

Given this view, some suggest there are opportunities to address this disparity. The Subcontinent or Indian Ocean or western Asia has been cited as one AOR that could merit a separate command. With long-term strategic emphasis on countering violent extremism in Afghanistan and Pakistan, it might be in the nation's best interest to establish a separate command rather than continuing to include them in USCENTCOM where the command's planners and decision makers must also focus on issues such as Iran's pursuit of nuclear weapons and regional influence, the Israel-Palestine impasse, and the fate of Syria and Egypt. India might also figure into this strategic recalculation as its relationships with Pakistan and China have a significant political-military impact in the region. Some believe India—currently the responsibility of USPACOM—might be a better fit under a separate Subcontinent COCOM.

Another area where a new COCOM could be warranted is Central Asia. Such a new command could include Turkmenistan, Kazakhstan, Kyrgyzstan, Tajikistan, and Uzbekistan—all presently under USCENTCOM—and could be a natural complement to a Subcontinent COCOM as many of the region's issues are more "localized" as tribes in the region tend not to conform to established political borders. These cross-border tribal and ethnic issues are viewed by many as key contributors to regional instability.

While the establishment of new COCOMs might have an academic appeal, critics note such a course of action might not be fiscally sustainable. Establishment of new COCOMs is viewed as being a resource-intensive undertaking—even if resources are taken from existing COCOMs. New COCOMs would require additional Joint-qualified senior and mid-level officers as well as supporting military, civilian, and possibly contractor staff. New COCOMs would also likely require additional physical infrastructure and if there is an intent to headquarter these new COCOMs in their AORs, there might also be political and diplomatic issues to consider. In a fiscally-constrained environment, these considerations might outweigh any operational benefits that might be derived from the establishment of new COCOMs.

Aside from Geographic COCOMs, there might also be cause to re-examine Functional COCOMs. One area for possible examination is if U.S. Cyber Command—currently a Subunified Command under USSTRATCOM—should be elevated to a full-fledged Combatant Command.[140] Proponents cite the following five benefits of this course of action:

- *Unity of Command/Effort:* The current DOD approach to cyberwarfare is scattered across the Services and defense agencies. The Services, the Defense Information Systems Agency (DISA), the National Security Agency (NSA), the Intelligence Community, and the other COCOMs have unsynchronized cyberspace warfighting capabilities. A separate and distinct USCYBERCOM would have greater authority, responsibility, legitimacy, and visibility than the current arrangement.
- *Synchronization:* Under the current command arrangement, USCYBERCOM might not have sufficient authority to fully synchronize cyber operations across the Services and COCOMs. This could lead to a situation in which a COCOM decides to conduct cyber operations within its AOR and, because there are no physical boundaries in cyber space, these actions could have an adverse impact on the other COCOMs.
- *Mass*: A central COCOM with exclusive authority could mass the cyber activities of the other COCOMs, Services, and DOD agencies into a coordinated effort to achieve mass effects on their intended target.
- *Offensive Operations*: One perceived benefit of elevating USCYBERCOM to full combatant command status is it would enable the U.S. government to place greater emphasis on offensive cyber operations as opposed to the current disparate emphasis on defensive cyber operations which are viewed by some as less effective than offensive operations.
- *Diverse Mission Focus:* Supporters argue the current command arrangement results in a lack of direction and discipline amongst DOD and government entities that makes oversight and, ultimately, funding more difficult. An elevated USCYBERCOM could serve to provide a single mission focus for the U.S. military.

While proponents suggest, in the long run, a separate USCYBERCOM would be cost efficient, such an undertaking in a fiscally-constrained environment could prove to be a difficult undertaking. There might also be resistance from the other COCOMs to ceding their cyber-related responsibilities to an elevated USCYBERCOM, arguing that they better understand the cyber threats in their specific regions than would a single entity responsible for a wider range of cyber threats. Because it might be difficult to identify, recruit, and retain cyber-qualified personnel that would likely be needed for a separate USCYBERCOM, cyber professionals from the COCOMs might be sought after to staff a new USCYBERCOM which could cause resentment from the COCOMs.

Is There an Alternative to COCOMs?

Some experts believe the COCOM construct is a relic of the Cold War where the central mission of the U.S. military was to prepare for and conduct combat operations in specific geographical regions against conventional armed forces. These experts suggest a more radical approach to the issue of COCOMs is required so they remain relevant in a post-September 11, 2001, world. Two possible alternatives to the current COCOM construct are usually discussed.

Replacing Subcomponent Commands with Joint Task Forces (JTFs)[141]

This proposal advocates retaining the COCOM headquarters and substituting a number of JTFs for the Service-centric Subcomponent Commands. This approach could streamline and reduce infrastructure and simplify command channels. These JTFs could be designed on a regional, functional basis, for specific operational tasks and would be more flexible and reduce response time to crises in the region.

Larger COCOM AORs might also benefit from several JTFs that could provide focused planning and operational execution in smaller, more manageable portions of their AORs. Another perceived benefit of this construct is if a specific condition within a COCOM's AOR is resolved, the JTF established to address the issue could be rapidly disestablished, thereby reducing personnel, infrastructure, and operations and maintenance (O&M) costs.

Replacing COCOMs with a Joint Interagency Organization[142]

This proposal advocates replacing COCOMs with permanent standing, civilian-led interagency organizations that would have regional responsibility for all aspects of U.S. foreign policy. These organizations would be led by highly-credentialed civilians, potentially with a four-star military deputy and would report directly to the President through the National Security Council (NSC).

These organizations would include representatives from all major U.S. government agencies, including DOD. This construct would change only the authority to integrate all elements of U.S. national power and DOD would continue to exercise its Title 10 authority by means of JIATFs.

A perceived benefit of this approach is it could result in a significant increase in unity of effort across all the instruments of U.S. national power through all phases of an operation to include pre and post-conflict. Another benefit is such an organization might better facilitate both coalition and alliance-based operations from a political standpoint as it may be more palatable for some nations to work with a civilian-led organization rather than a military-centric one. This new construct might also have benefits for both regional engagement and developmental efforts thereby reducing the military "face" of these operations, particularly in regions that are sensitive to U.S. military presence.

These are but two possible alternatives to the current COCOM construct. While Congress might not choose to directly address COCOM reform (particularly if there is a general belief the current COCOM construct meets current and future security needs) it is possible that, as Congress, the Administration, and DOD continue to pursue government-wide efficiencies aimed at reducing federal spending, these and other alternatives might inform the debate.

Appendix. 2011 UCP COCOM Areas of Responsibility

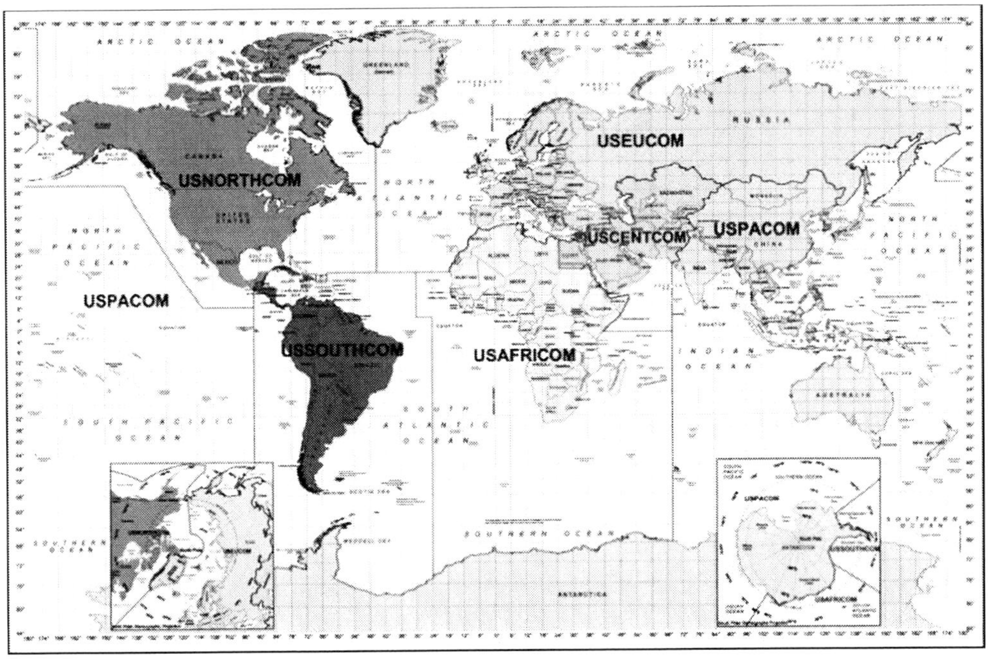

End Notes

[1] Joint Publication 1-02, Department of Defense Dictionary of Military and Associated Terms, November 8, 2010 (as amended through May 15, 2011), p. 385.

[2] Information in this section is from a briefing provided to CRS by the Joint Staff Plans Division on October 7, 2010. Unless otherwise noted, information in this section is taken from the Office of the Chairman of the Joint Chiefs of Staff Joint History Office, "The History of the Unified Command Plan 1946-1999," Washington, DC, 2003.

[3] Joint Publication 1-02 defines a unified command as a "command with a broad continuing mission under a single commander and composed of significant assigned components of two or more Military Departments that is established and so designated by the President, through the Secretary of Defense with the advice and assistance of the Chairman of the Joint Chiefs of Staff."

[4] Joint Publication 1-02 defines a specified command as "a command that has a broad, continuing mission, normally functional, and is established and so designated by the President through the Secretary of Defense with the advice and assistance of the Chairman of the Joint Chiefs of Staff. It normally is composed of forces from a single Military Department."

[5] Ibid., p. 60.

[6] Cynthia A. Watson, "Combatant Commands: Origins, Structure, and Engagement," Praeger Security International, 2011, p. 15 (hereinafter, Cynthia A. Watson).

[7] Unless otherwise noted, information in this section is taken from the Office of the Chairman of the Joint Chiefs of Staff Joint History Office, "The History of the Unified Command Plan 1946-1999," Washington, DC, 2003.

[8] Ibid., p. 11.

[9] Charles A. Stevenson, "Underlying Assumptions of the National Security Act of 1947," *Joint Forces Quarterly*, 1st Quarter 2008.

[10] Office of the Chairman of the Joint Chiefs of Staff Joint History Office, "The History of the Unified Command Plan 1946-1999," Washington, DC, 2003, p 11.

[11] Cynthia A. Watson, p. 13.

[12] "The History of the Unified Command Plan 1946-1999," p. 15.

[13] Ibid., p. 26.
[14] Cynthia A. Watson, pp. 14-15.
[15] Ibid.
[16] Information in this section is from a briefing provided to CRS by the Joint Staff Plans Division on October 7, 2010.
[17] SIPRNET stands for Secret Internet Protocol Router Network DOD's classified computer network.
[18] Tank is the euphemism for the Joint Chiefs of Staff Conference Room in the Pentagon where sensitive important senior-level meetings are held.
[19] Information in this section is from a briefing provided to CRS by the Joint Staff Plans Division on October 7, 2010.
[20] Ibid.
[21] U.S. Department of Defense News Release, DOD Releases Unified Command Plan 2011, N. 288-11, April 8, 2011.
[22] Taken from information provided to CRS by the Senate Armed Services Committee on September 13, 2011.
[23] From Joint Publication 1-02, Department of Defense Dictionary of Military and Associated Terms, as amended through September 15, 2011, a Standing Joint Force Headquarters or SJFHQ is a staff organization operating under a flag or general officer providing a combatant commander with a full-time, trained joint command and control element integrated into the combatant commander's staff whose focus is on contingency and crisis action planning.
[24] Cynthia A. Watson, p. 11.
[25] Ibid.
[26] Unless otherwise noted, information in this section is taken from Sections 161-168, Title 10–Armed Forces, pp. 144- 154, and Cynthia A. Watson, pp. 2-6.
[27] Information in this section is taken from Cynthia A. Watson, pp. 2-6.
[28] Ibid., pp 19-21.
[29] Information in this section is taken from Department of Defense Report to Congress Pursuant to Section 944 of the Ike Skelton National Defense Authorization Act for Fiscal Year 2011 P.L. 111-383, Department of Defense Report on Organizational Structures of the Headquarters of the Geographic combatant Commands, June 2011, p. 3.
[30] From USJFCOM website, http://www.jfcom.mil/about/priorities.htm.
[31] USJFCOM Facebook page, http://www.facebook.com/pages/JFCOM/111646735521012.
[32] DOD News Release, "Secretary Gates Announces Efficiencies Initiatives," No. 706-10, August 9, 2010.
[33] Ibid.
[34] Bill Bartel, "Jobs to be Lost in the Region: 1,900; Annual Savings for the Military: $400M; Date of JFCOM Demise: By Nov.," *Norfolk-Virginian Pilot,* January 11, 2011.
[35] Ann Roosevelt, "JFCOM Functions to Gather Under Joint Staff," *Defense Daily,* February 11, 2011.
[36] Information in this section is taken from Statement of General Raymond T. Odierno, U.S. Army, Commander, United States Joint Forces Command, Improving the Readiness of U.S. Forces Through Military Jointness, HASC Readiness Subcommittee, March 31, 2011 and "DOD Approves Reassignment of Certain JFCOM Activities, *Defense Daily,* May 3, 2011.
[37] For detailed information on USSOCOM, subcomponents, and assigned forces see CRS Report RS21048, *U.S. Special Operations Forces (SOF): Background and Issues for Congress*, by Andrew Feickert.
[38] Posture Statement of Admiral Eric T. Olson, USN, Commander United States Special Operations Command before the 112th Congress Senate Armed Services Committee, March 1, 2011.
[39] Information in this section is taken from author's National War College course materials, USSOCOM History Book, "Founding and Evolution of USSOCOM," pp. 5-11.
[40] Information from this section is taken from CRS Report RS21048, *U.S. Special Operations Forces (SOF): Background and Issues for Congress*, by Andrew Feickert and USSOCOM website, http://www.socom.mil.
[41] Information in this section is taken from the JSOC web page, http://www.socom.mil/Pages/JointSpecialOperationsCommand.aspx.
[42] Jennifer D. Kibbe, "The Rise of the Shadow Warriors," Foreign Affairs, Volume 83, Number 2, March/April 2004 and Sean D. Naylor, "JSOC to Become Three-Star Command," *Army Times*, February 13, 2006; and U.S. Army Special Operations Command News Service, "Gates Nominates McRaven, Thurman for Senior Posts," Release Number: 110303-02, March 3, 2011, http://www.soc.mil/UNS/Releases/2011/March/110303-02.html.
[43] Information in this section is taken from the JSOU web page, https://jsou.socom.mil.

[44] Information in this section is taken from Special Operations Command–Joint Capabilities website, http://www.sojc.socom.mil.

[45] CQ Congressional Transcripts, “”House Armed Services Committee Holds Hearing on the Proposed Fiscal Year 2012 Budget for the Defense Department’s U.S. Central Command and U.S. Special Operations Command,” March 3, 2011, p. 4.

[46] Ibid., p. 5.

[47] Information in this section is taken from USSTRATCOM’s web page, http://www. stratcom.mil, and Statement of General C. Robert Kehler, Commander, United States Strategic Command before the Senate Committee on Armed Services, March 29, 2011.

[48] http://www.stratcom.mil/history/.

[49] http://www.stratcom.mil/service_components/.

[50] http://www.stratcom.mil/functional_components/, and U.S. Strategic Command “Snap Shot,”–A Summary of USSTRATCOM Facts and Information, March 2011.

[51] Congressional Transcripts, “Senate Armed Services Committee Holds Hearing on the Defense Authorization Request for Fiscal Year 2012 and Future Years for the U.S. European Command and U.S. Strategic Command,” March 29, 2011.

[52] Statement of General C. Robert Kehler, Commander, United States Strategic Command before the Senate Committee on Armed Services, March 29, 2011.

[53] For additional information on U.S. nuclear weapons see CRS Report RL30345, *U.S. Nuclear Weapons: Policy, Force Structure, and Arms Control Issues*, by Amy F. Woolf.

[54] For additional information on USCYBERCOM see CRS Report R40836, *Cybersecurity: Current Legislation, Executive Branch Initiatives, and Options for Congress*, by Catherine A. Theohary and John Rollins

[55] Statement of General C. Robert Kehler, Commander, United States Strategic Command before the Senate Committee on Armed Services, March 29, 2011, pp. 15-16.

[56] Information in this section is taken from http://www.transcom.mil/about/whatIs.cfm.

[57] Taken directly from http://www.transcom.mil/about/briefHistory.cfm.

[58] http://www.transcom.mil/about/cocom.cfm.

[59] Statement of General Duncan J. McNabb, USAF, Commander, United States Transportation Command Before the Senate Armed Services Committee on the State of the Command, April 7, 2011.

[60] Ibid.

[61] Jason Sherman, “TRANSCOM Awaits Nod from White House on Expanded Portfolio (Update),” *InsideDefense.com,* April 6, 2011.

[62] Ibid.

[63] For additional information on USAFRICOM see CRS Report RL34003, *Africa Command: U.S. Strategic Interests and the Role of the U.S. Military in Africa*, by Lauren Ploch.

[64] Fact Sheet: United States Africa Command, http://www.africom.mil/getArticle.asp?art=1644.

[65] AFRICOM FAQs, http://www.africom.mil/AfricomFAQ.asp.

[66] Cynthia A. Watson, p. 174.

[67] Information in this section is taken from Milady Ortiz, “U.S. Africa Command: A New Way of Thinking,” National Security Watch, Association of the U.S. Army Institute of Land Warfare,” March 13, 2008.

[68] Cynthia A. Watson, p. 174.

[69] For additional information on regional terrorism see CRS Report R41473, *Countering Terrorism in East Africa: The U.S. Response*, by Lauren Ploch.

[70] For additional information on piracy around the Horn of Africa see CRS Report R40528, *Piracy off the Horn of Africa*, by Lauren Ploch et al.

[71] http://www.usaraf.army.mil/.

[72] http://www.naveur-navaf.navy.mil/.

[73] http://www.africom.mil/pdfFiles/AFRICOMfactsheet05AUG2011.pdf.

[74] Ibid.

[75] AFRICOM FAQs, http://www.africom.mil/AfricomFAQ.asp.

[76] USSOCOM 2011 Fact Book, http://www.socom.mil/News/Documents/ USSOCOMFact Book_2011.pdf.

[77] Statement: AFRICOM Commander on Commencement of Military Strikes in Libya, March 19, 2011, and Jim Garamone, “NATO Assumes Command of Libya Operations,” American Forces Press Service, March 31, 2011.

[78] For additional information on military operations against Libya see CRS Report R41725, *Operation Odyssey Dawn (Libya): Background and Issues for Congress*, coordinated by Jeremiah Gertler.

[79] U.S. Department of States, "Transcript: NATO Representative Announces End of Operation in Libya," October 24, 2011.

[80] Information in this section is taken from The White House Office of the Press Secretary to the speaker of the House of Representatives and the President Pro Tempore of the Senate, October 14, 2011; and Lisa Daniels, "Official: Troops in Central Africa for Months, Not Years," *American Forces Press Service,* October 26, 2011.

[81] USAFRICOM, OEF Trans-Sahara, http://www.africom.mil/fetchBinary. asp?pdfID= 20100 526130828.

[82] Fact Sheet: USAFRICOM 2011 Exercises, April 25, 2011.

[83] Statement of General Carter F. Ham, USA, Commander United States Africa Command Before the House Armed Services Committee, April 5, 2011.

[84] Ibid., p. 26.

[85] Ibid., p 27.

[86] http://www.centcom.mil.

[87] http://www.centcom.mil/en/about-centcom/our-history/.

[88] Cynthia A. Watson, p. 142.

[89] Ibid., pp. 141-142.

[90] http://www.afcent.af.mil/main/welcome.asp.

[91] http://www.marines.mil/unit/marforcent/Pages/MARCENT_Main.aspx.

[92] http://www.socom.mil/News/Documents/USSOCOMFactBook2011.pdf.

[93] Statement of U.S. Marine Corps General James N. Mattis, U.S. Central Command Commander, before the Senate Armed Services Committee on March 1, 2011, about the posture of U.S. Central Command.

[94] For detailed information on operations in Iraq and Afghanistan see CRS Report RL34387, *Operation Iraqi Freedom: Strategies, Approaches, Results, and Issues for Congress*, by Catherine Dale and CRS Report R40156, *War in Afghanistan: Strategy, Operations, and Issues for Congress*, by Catherine Dale.

[95] CQ Congressional Transcripts, "Senate Armed Services Committee Holds Hearing on the Proposed Fiscal Year 2012 Budget for the Defense Department's U.S. Central Command and U.S. Special Operations Command," March 1, 2011.

[96] http://www.eucom.mil/documents/fact-sheets/EUCOM-Factsheet-Command-2010.pdf.

[97] http://www.nato.int/cps/en/natolive/topics_50110.htm.

[98] http://www.eucom.mil/documents/fact-sheets/EUCOM-Factsheet-History-2010.pdf.

[99] http://www.eucom.mil/documents/fact-sheets/EUCOM-Factsheet-Command-2010.pdf and House and Senate Armed Services Committees Testimony of Admiral James G. Stavridis, United States Navy, Commander United States European Command Before the 112th Congress, 2011, March 2011.

[100] House and Senate Armed Services Committees Testimony of Admiral James G. Stavridis, United States Navy, Commander United States European Command Before the 112th Congress, 2011, March 2011.

[101] For more information on NATO in Afghanistan see CRS Report RL33627, *NATO in Afghanistan: A Test of the Transatlantic Alliance*, by Vincent Morelli and Paul Belkin.

[102] House and Senate Armed Services Committees Testimony of Admiral James G. Stavridis, United States Navy, Commander United States European Command Before the 112th Congress, 2011, March 2011; and CQ Congressional Transcripts, "Senate Armed Services Committee Holdings Hearing on the Defense Authorization Request for Fiscal 2012 and Future Years for the U.S. European Command and U.S. Strategic Command," March 29, 2011.

[103] For additional information see CRS Report RL34051, *Long-Range Ballistic Missile Defense in Europe*, by Steven A. Hildreth and Carl Ek.

[104] Tom Shanker, "U.S. Hails Deal with Turkey on Missile Defense," *New York Times,* September 15, 2011.

[105] http://www.northcom.mil/About/index.html.

[106] Ibid. and http://www.norad.mil/about/index.html.

[107] Ibid.

[108] For a definition of the Posse Comitatus Act that limits the use of military force in law enforcement, see http://www.northcom.mil/About/history_education/posse.html.

[109] For additional information on Posse Comitatus, see CRS Report RS22266, *The Use of Federal Troops for Disaster Assistance: Legal Issues*, by Jennifer K. Elsea and R. Chuck Mason.

[110] http://www.northcom.mil/About/history_education/history.html.

[111] http://www.northcom.mil/About/index.html.

[112] CQ Transcripts, "Senate Armed Services Committee Holds Hearing on the Proposed Defense Authorization Request for Fiscal 2012 and Future Years for the U.S. Northern Command and U.S. Southern Command," April 5, 2011.

[113] Ibid.

[114] For additional information on Mexican operations against TCOs, see CRS Report R41576, *Mexico's Drug Trafficking Organizations: Source and Scope of the Rising Violence*, by June S. Beittel.

[115] For additional information on the Arctic see CRS Report R41153, *Changes in the Arctic: Background and Issues for Congress*, coordinated by Ronald O'Rourke

[116] http://www.pacom.mil/web/Site_Pages/USPACOM/Facts.shtml.

[117] http://www.pacom.mil/web/site_pages/uspacom/history.shtm.

[118] Cynthia A. Watson, pp. 49-55.

[119] http://8tharmy.korea.army.mil/mission.asp.

[120] http://www.cpf.navy.mil/about/facts/.

[121] Statement of Admiral Robert F. Willard, U.S. Navy, Commander, U.S. Pacific Command Before the Senate Armed Services Committee on U.S. Pacific Command Posture, April 12, 2011, pp. 27-29.

[122] Ibid., p. 3.

[123] For additional information on North Korean nuclear weapons see CRS Report RL34256, *North Korea's Nuclear Weapons: Technical Issues*, by Mary Beth Nikitin.

[124] For information on Chinese naval modernization see CRS Report RL33153, *China Naval Modernization: Implications for U.S. Navy Capabilities—Background and Issues for Congress*, by Ronald O'Rourke.

[125] http://www.southcom.mil/AppsSC/pages/about.php.

[126] http://www.southcom.mil/AppsSC/factfiles.php?id=76.

[127] http://www.southcom.mil/AppsSC/pages/team.php.

[128] Ibid.

[129] Posture Statement of General Douglas M. Fraser, United States Air Force, Commander, United States Southern Command, Before the 112th Congress, Senate Armed Services Committee, April 5, 2011.

[130] Ibid.

[131] For additional information on U.S. Interagency issues see CRS Report RL34455, *Organizing the U.S. Government for National Security: Overview of the Interagency Reform Debates*, by Catherine Dale, Nina M. Serafino, and Pat Towell.

[132] Edward Marks, "Rethinking the Geographic Combatant Commands, *InterAgency Journal,* Vol. 1, Issue 1, Fall 2010, p. 20.

[133] Dana Priest, "A Four-Star Foreign Policy? U.S. Commanders Wield Rising Clout, Autonomy," September 28, 2000; "An Engagement in 10 Time Zones: Zinni Crosses Central Asia, Holding Hands, Building Trust," September 29, 2000; "Standing Up to State and Congress," September 30, 2000, *Washington Post.*

[134] Testimony from Mark Malan, Refugees International, Washington, DC, "AFRICOM: A Wolf in Sheep's Clothing?" before the Subcommittee on African Affairs, Committee on Foreign Relations, U.S. Senate, at a hearing entitled "Exploring the U.S. Africa Command and a New Strategic Relationship with Africa," August 1, 2007.

[135] Howard D. Belote, "Proconsuls, Pretenders, or Professionals? The Political Role of Regional Combatant Commanders," Chairman of the Joint Chiefs of Staff Strategy Essay Competition, 2004.

[136] Ibid., p. 8.

[137] Unless otherwise noted, information in this section is taken from Cynthia A. Watson, "Combatant Commands: Origins, Structure, and Engagement," Praeger Security International, 2011, pp. 185-186.

[138] Edward Marks, p. 21.

[139] Ibid.

[140] Information in this section, unless otherwise noted, is from David M. Hollis, "USCYBERCOM: The Need for a Combatant Command Versus a Subunified Command," *Joint Forces Quarterly,* Issue 58, 3rd Quarter 2010, pp. 48-53

[141] Information in this section, unless otherwise noted, is taken from W. Spencer Johnson, "New Challengers for the Unified Command Plan," *Joint Forces Quarterly,* Summer 2002, pp. 62-70.

[142] Brigadier General Jeffery Buchanan, U.S. Army, Captain Maxie Y. Davis, U.S. Navy, and Colonel Lee Wright, U.S. Air Force, "Death of the Combatant Command? Toward a Joint Interagency Approach," *Joint Forces Quarterly,* Issue 52, 1st Quarter 2009.

In: Combatant Commands in the United States Armed Forces ISBN: 978-1-61942-924-6
Editors: Quincy N. Butler and Harris A. Simmons

Chapter 2

STATEMENT OF GENERAL CARTER F. HAM, USA COMMANDER UNITED STATES AFRICA COMMAND BEFORE THE SENATE ARMED SERVICES COMMITTEE*

7 April 2011

INTRODUCTION

It is my privilege as Commander of United States Africa Command to present to Congress our Posture Statement for 2011. United States Africa Command protects American lives and promotes our national interests by advancing security and stability in Africa. We follow two main lines of effort building the security capacity of our African partners and preparing for a wide range of potential crises. U.S. Africa Command's operations, programs, and activities contribute to reducing the threats to our citizens and interests both abroad and at home by helping African states provide for their own security. We seek to enhance regional stability through support to and partnership with African regional organizations. Our planning and training are designed to prevent conflict while simultaneously ensuring that U.S. Africa Command is prepared to respond decisively to any crisis when the President so directs.

A prosperous and stable Africa is strategically important to the United States. An Africa that can generate and sustain broad based economic development will contribute to global growth, which is a long-standing American interest. However, poverty in many parts of Africa contributes to an insidious cycle of instability, conflict, environmental degradation, and disease that erodes confidence in national institutions and governing capacity. This in turn often creates the conditions for the emergence of a wide-range of transnational security threats that can threaten the American homeland and our regional interests.

---

* These remarks were delivered as testimony given on 7 April 2011. General Carter F. Ham, USA Commander United States Africa Command before the Senate Armed Services Committee.

Our primary effort for increasing stability and deterring conflict is focused on building partner capacity one of six core Department of Defense (DOD) mission areas identified in the 2010 Quadrennial Defense Review (QDR). The Command is helping African states transform their militaries into operationally capable and professional institutions that are subordinate to civilian authority, respect human rights, adhere to the rule of law, and are viewed by their citizens as servants and protectors of the people. We assist our African partners in building capacities to counter transnational threats from violent extremist organizations; to stem illicit trafficking in humans, narcotics, and weapons; to support peacekeeping operations; and to address the consequences of humanitarian disasters whether man-made or natural that cause loss of life and displace populations. In many instances, the positive effects we achieve are disproportionate to the modest investment in resources.

Dramatic events taking place in Africa, as demonstrated by the unfolding situation across North Africa, have shown how quickly the strategic environment can change. It has also shown the value of the Nation's investment in the Command. As of the writing of this report, the situation in Libya remains unclear. The Command is prepared to respond in a variety of ways pending National decisions. We will maintain our steady focus on security cooperation with our African partners, and stand ready to protect American lives and interests.

Strategic Environment

The challenges and opportunities in U.S. Africa Command's Area of Responsibility (AOR) are dynamic and complex. Our AOR includes 52 African States more than 800 ethnic groups, over 1000 languages, and a diverse geography three and a half times the size of the continental United States.[1] Its rapidly growing population presents a complex mix of political, economic, social, and demographic challenges.

Continent-wide Trends

The global economic crisis disrupted a period of sustained economic growth in Africa. However, according to a 2010 report developed jointly by the African Development Bank, the Organization for Economic Cooperation and Development, and the United Nations (UN) Economic Commission on Africa, the continent's economies are resilient and the prospects for regaining economic momentum are good. Earlier debt relief and aid programs ensured that many African governments were not burdened with onerous financial obligations, and their commodity-based economies appear poised to rebound faster than countries that are more dependent on financial and manufacturing sectors.

Demographic trends highlight the urgent need for sustained and broad-based economic growth throughout Africa. High birth rates, a youth bulge, and urban congestion stress the ability of many governments to provide basic services, education and jobs. Forty-three percent of sub-Saharan Africa's population is below the age of 15. This population could provide further advances in developing and consolidating democracy; however, if unaddressed, this potential pool of undereducated and unemployed youth could present a possible source of instability and potential recruiting pool for violent extremist organizations or narcotics traffickers. The rapid expansion of already strained urban areas with limited infrastructure will increase competition for limited jobs, social services, housing, food, and water as Africa's population is expected to double to two billion by 2050.

A welcome development across the continent is that several African conflicts have ended in the last ten years—Liberia, Angola, Burundi, Sierra Leone, Chad, and the North-South Sudanese conflict. However, as recent events in Cote d'Ivoire and Darfur illustrate, many of these settlements and compromises remain fragile.

African states are working hard to meet their own security challenges. Today, Africans comprise 30 percent of UN peacekeeping forces worldwide and 50 percent of peacekeeping forces deployed across the African continent. At any given time there are approximately 25,000 African soldiers deployed in support of UN and African Union (AU) peacekeeping missions. Five African countries—Nigeria, Rwanda, Ghana, Ethiopia, and South Africa—rank amongst the top 15 UN troop-contributing nations, and Uganda and Burundi are the major contributors to the AU peacekeeping mission in Somalia. Also, the development of the AU and its Regional Economic Communities, growing regional cooperation among neighboring states, the efforts to establish the African Standby Force, regional cooperation by Gulf of Guinea states to improve maritime security, and combined efforts to combat the Lord's Resistance Army (LRA) demonstrate a growing continent-wide commitment to establishing stability.

Though much of Africa is peaceful today, conflict remains a significant problem that drains resources and saps the confidence needed for Africa to realize its potential. These lingering conflicts are predominantly driven by internal fissures rather than external interstate disputes. The conflict between the Nigerian federal government and a variety of alienated groups in the oil and gas rich Niger Delta region continues and is a concern as the 2011 elections approach. The Democratic Republic of the Congo (DRC) continues to recover slowly from one of Africa's most devastating wars. With over 20,000 soldiers, the UN mission in the DRC is the largest and most expensive peacekeeping operation in the world. Additionally, the LRA, although diminished since Uganda began operations against it in 2008, continues its horrific actions, which threaten stability in central Africa.

Sudan remains a concern as the governments of the north and south implement the results of the January 2011 independence referendum. North and South Sudan will need to resolve contentious issues like oil wealth sharing, debt, and border demarcation before the end of the mandated Comprehensive Peace Agreement timeline set for July 2011. The new government of Southern Sudan will face significant challenges uniting an ethnically disparate populous while building both state institutions and economic infrastructure.

Concerning elections, the positive democratizing trend that emerged in the 1990s has slowed somewhat during the last few years as some nations continue to suffer from constitutional tampering, rigged elections, and intimidation of opposition candidates. African elections in the near term are likely to produce mixed results for democratization. However, the proliferation of civil and political opposition groups enabled by social media, increased voter participation, local press coverage of political events, greater citizen access to information, and continued international attention will help promote transparent and democratic elections in the 22 national contests scheduled in 2011. The recent changes of government in Tunisia and Egypt vividly illustrate these trends, as the power of social media and greater citizen access to information has empowered ordinary citizens.

Good governance builds the trust and confidence necessary for regional cooperation and economic development, and provides the leverage for national and regional programs to foster growth and stability. The African states making progress toward free and fair elections, open regulatory practices, and the provision of essential services are, more often than not, reliable partners trusted by their neighbors. Those states with weak governments are less stable and less dependable.

Transnational Threats

There are numerous transnational threats in Africa, with violent extremism, piracy, and narcotics trafficking constituting three of the most dangerous. Many African states recognize these threats and are taking positive actions—often with Africa Command's assistance—to confront them.

Somalia remains a failed state: divided, weak, and fragile. Despite the intentions of the Transitional Federal Government (TFG) to establish the sinews of a functioning state, Somalia is unable to provide essential services or control of its territory on its own. It does not have a civil service, central bank, judicial system, police force, or functioning cross-clan military. Inter-clan rivalries and the Islamic extremist groups al-Shabaab and Hizbul al-Islam continue to challenge Somali's ability to form a functioning and stable government, and al-Shabaab controls much of southern Somalia. The survival of the TFG in Mogadishu depends, in large measure, on the presence of the African Union Mission in Somalia (AMISOM) and the more than 8,000 troops supplied by willing African partners.

Linked to Somalia's instability is al-Qaida's dramatic increase in influence in east Africa over the last year. In early 2010, al-Shabaab announced their alignment with al-Qaida. This alliance provides al-Qaida a safe haven to plan global terror operations, train foreign fighters, and conduct global terror operations. The July 2010 attacks in Kampala, Uganda, demonstrate a willingness and capability to expand the conflict beyond Somalia. This situation poses a direct threat to the security of the United States.

Piracy remains a significant problem off the coast of Somalia. The murder of four U.S. citizens aboard the sailing vessel *Quest* this February demonstrates that piracy is a lethal threat— not just an economic inconvenience or acceptable business risk. The pirates have adapted to the increased international naval presence in the Gulf of Aden by extending their operational reach as far south as the Mozambique Channel and east into the Indian Ocean by hundreds of miles. The number of ships and hostages held by Somalia-based pirates is at an all-time high. The increasing operational reach of Somali pirates vividly illustrates that the dangers emanating from ungoverned spaces rarely remain local.

The Sahel region continues to experience attacks and kidnappings by al-Qaida in the Lands of the Islamic Maghreb, an Algerian-based group with relative safe haven in northern Mali.

If al-Qaida affiliates grow unchecked in the Horn of Africa or across the Sahel, it may lead to further attacks against U.S. interests overseas or in the homeland.

Illicit trafficking of narcotics poses a significant threat to regional stability in both West and East Africa. Western Africa is a critical transshipment point for South American cocaine.

Increasing European cocaine demand, transportation availability, and negligible risk of interdiction contribute to West Africa's importance in the narcotics trade.

The cocaine transiting through this area constitutes from 30 to 60 percent of European demand, with relative wholesale values exceeding the gross domestic products of some affected West African states. East Africa is also facing trafficking challenges, but in the form of Afghan heroin via the Makran Coast of Pakistan pouring into world markets.

UNITED STATES AFRICA COMMAND'S APPROACH

U.S. Africa Command's approach to the continent is guided by our national interests and priorities, our theater objectives, and three operating principles.

- We listen and learn from our African partners;
- We understand and address the continent using a regional framework; and,
- We collaborate as part of an interagency team.

U.S. Security Interests, Priorities, and Theater Objectives

U.S. Africa Command's programs and activities directly support American national security interests. Our vital national security interest in Africa is protecting the lives and interests of the American people by reducing threats to the homeland and abroad.

We support the United States Government's (USG) five priorities in Africa: good governance, economic progress, preventing and resolving conflicts, strong public health programs, and helping our African partners develop the capacity to meet the demands of transnational challenges. In supporting these national priorities, U.S. Africa Command focuses on preventing and resolving conflict and helping our African partners develop their own security capacity.

To focus our efforts, U.S. Africa Command has established the following theater objectives:

- Ensure that the al-Qaida networks and associated violent extremists do not attack the United States;
- Maintain assured access and freedom of movement throughout our AOR;
- Assist African states and regional organizations in developing the will, capability, and capacity to combat transnational threats such as terrorism, piracy, and the illicit trafficking of weapons, people and narcotics;
- Assist African states and regional organizations in developing the capacity to execute effective continental peace operations and to respond to crises; and,
- Encourage African militaries to operate under civilian authority, respect the rule of law, abide by international human rights norms, and contribute to stability in their respective states.

Listening to Our African Partners

U.S. Africa Command values the views of our African partners and learns from them. Our partners have expressed four common defense-oriented goals consistent with American interests and the Command's theater objectives. These goals are:

- That they have capable and accountable military forces that perform professionally and with integrity;
- That their forces are supported and sustained by effective, legitimate, and professional security institutions;
- That they have the capability to exercise the means nationally and regionally to dissuade, deter, and defeat transnational threats; and,
- That they have the capacity to increase their support to international peacekeeping efforts.

These goals are in agreement with our emphasis on preventing conflict and enhancing regional stability to protect American lives and interests in Africa and in our homeland.

A Regional Perspective

To identify and analyze the relationships among crosscutting security issues on the vast continent, we view Africa regionally. The five regions—East Africa, West Africa, Central Africa, North Africa, and Southern Africa roughly correspond to the AU's five primary Regional Economic Communities. The Command's staff is responsible for establishing and supporting our African partnerships and developing programs and activities, and is organized along the same regional lines.

East Africa

The East African states include Djibouti, Eritrea, Ethiopia, Kenya, Rwanda, Somalia, Sudan, Tanzania, Burundi, Uganda, and the island states of Comoros and Seychelles. The interlocking security challenges of Somalia's instability, Southern Sudan's transition to statehood, al-Shabaab's dangerous alignment with al-Qaida, which threatens not only the region but also the American homeland and interests, and the persistent threat from LRA require both regional and bilateral solutions. East African states recognize these challenges and are taking steps to meet them. Uganda is a major troop contributor to AMISOM and is involved in operations against the LRA. Moreover, Djibouti has emerged as a stable partner in a dangerous region and provides our only enduring military infrastructure in continental Africa, Camp Lemonnier

In East Africa, Combined Joint Task Force – Horn of Africa (CJTF-HOA) is essential to U.S. Africa Command's effort to build the partner capacity necessary to counter violent extremism and address regional security issues. It is the Command's element for conducting Operation Enduring Freedom – Horn of Africa. CJTF-HOA conducts engagement activities by employing small teams throughout the region. Typical military-to-military engagements include officer and non-commissioned officer mentoring, counter-terrorism training, peace support operations, and disaster response planning and preparation. Civil Affairs teams operating from Camp Lemonnier in Djibouti help partner nations improve civil-military relations in their communities. CJTF-HOA is a model for multinational and interagency collaboration, and plays a key role in supporting two important Department of State initiatives; the Africa Contingency Operations Training and Assistance (ACOTA) program as part of the larger Global Peace Operations Initiative, and the Partnership for Regional East Africa Counterterrorism (PREACT).

West Africa

The countries of Senegal, Mali, Gambia, Guinea Bissau, Guinea, Sierra Leone, Liberia, Cote d'Ivoire, Burkina Faso, Ghana, Togo, Benin, Nigeria, Niger, Cape Verde, and Sao Tome and Principe comprise West Africa. West Africa features a diverse population, straddles the transition zone between Christianity and Islam, and retains overlapping French, Portuguese, and British influences. Today West Africa experiences political instability and great disparities in wealth. Recent election disputes in Guinea and Cote d'Ivoire are indicative of stubborn regional problems. Liberia's revitalization, however, demonstrates that West African states can take positive steps to overcome violence, poverty, and disorder.

Nigeria, by virtue of its location, population, significant gross domestic product, and oil reserves is a major power in the region, and one of the continent's most politically and economically significant states. Nigeria provides regional leadership through the Gulf of Guinea Commission and Economic Community of West Africa States. Nigeria, along with Ghana and Senegal, are emerging in West Africa as critical partners essential to our efforts to enhance stability and security in this important region.

U.S. Africa Command assists in developing the capacity of individual West African states and encourages regional security cooperation. Exercise FLINTLOCK 11 in February and March 2011 was a special operations exercise focused on military interoperability and capacity building with partner nations in North and West Africa. Approximately 669 African, European and U.S. participants from 17 nations trained together in this 17-day exercise. In addition, 9 of the 15 West African countries participate in Africa Partnership Station, where their focus is on enhancing their maritime capabilities and developing multinational approaches to security in the Gulf of Guinea.

In support of Liberia's revitalization, U.S. Africa Command contributes to a sustained multi-year defense sector reform operation to help establish effective armed forces in Liberia: Operation ONWARD LIBERTY (OOL). OOL is a dedicated DOD and Department of State effort that follows three interrelated lines of operation that include security force assistance to the Armed Forces of Liberia, a U.S. Coast Guard effort to help establish a maritime law enforcement capability in the Liberian Economic Exclusion Zone, and mentorship and advisor support to the Liberian Ministry of Defense.

Central Africa

The Central African states of Chad, Cameroon, Equatorial Guinea, Gabon, Republic of the Congo, Angola, DRC, and Central African Republic extend from the middle of the continent east to the Great Lakes, and westward to the Atlantic Ocean. This is one of the most geographically diverse regions in the world and is wealthy in resources. For example, Angola is reaping the rewards of new oil wealth, but is still developing programs to generate broad-based, sustained economic growth. Corruption, weak political institutions, high instances of pandemic disease, and an inability to secure their maritime domain, borders, and natural resources hinder Central Africa's progress.

The lingering effects of the DRC conflict remain the primary challenge to peace and stability in the region. As part of the effort to help the DRC establish the rule of law and legitimate national institutions, U.S. Africa Command has been working with other U.S. agencies and international partners to improve the Congolese military's ability to serve its citizens. Specific examples of U.S. government efforts include strengthening the military justice system, improving training capacity, developing agricultural projects designed to reduce military personnel dependency on local populations, and improving indigenous training capacity.

The LRA continues to threaten vulnerable populations in northern Uganda and parts of the eastern DRC, Central African Republic, and Southern Sudan. The "Lord's Resistance Army Disarmament and Northern Uganda Reconstruction Act of 2009" directed the United States Government to develop a strategy to "mitigate and eliminate the threat to civilians and regional stability" posed by the LRA. U.S. Africa Command is the key DOD participant in the interagency effort to assist our regional partners in apprehending or removing the group's leaders, to protect civilians, to provide humanitarian assistance, and to promote justice and reconciliation in LRA-affected areas.

North Africa

The countries of North Africa are Algeria, Libya, Mauritania, Morocco, and Tunisia. Separated physically and culturally from sub-Saharan Africa by the Sahara Desert, this region maintains strong links to Southern Europe and the Middle East. North Africa is a critical gateway to sea-lane choke points as well as air and sea routes into Europe and sub-Saharan Africa. Trafficking of all types (human, drugs, weapons), economic stagnation, and ungoverned space in the Sahel make the region vulnerable to extremist influences and activities. The political and social upheaval that erupted across North Africa has captured the attention of the international community. However, it is too early to assess properly either the myriad causes or the outcomes of these important events.

Al-Qaida in the Lands of the Islamic Maghreb has emerged as a direct threat to the stability and security of North Africa as well as the West African Sahelian countries of Niger and Mali. To assist these countries in meeting this extremist challenge, U.S. Africa Command supports the Department of State led interagency Trans-Sahara Counter-Terrorism Partnership through Operation ENDURING FREEDOM – Trans-Sahara (OEF-TS). OEF-TS strengthens regional counterterrorism and security capabilities by employing small Mobile Training Teams, Civil Military Support Elements, Joint Planning and Assistance Teams, and through Senior Leader Engagements.

Southern Africa

Southern Africa comprises nine continental states including Namibia, South Africa, Lesotho, Swaziland, Botswana, Mozambique, Zimbabwe, Zambia, Malawi and two island states—Madagascar and Mauritius. In many dimensions, Southern Africa is highly developed economically; yet, the region still has some significant problems. Governance challenges, inflation, and refugees from Zimbabwe present challenges to Southern African nations. Two states with great potential, Zimbabwe and Madagascar, have difficult internal political challenges and are under economic sanctions, while Botswana continues to rise from one of the world's poorest countries to middle-income status. South Africa and the United States recently institutionalized a steadily improving era of cooperation when Secretary Clinton and South Africa's Minister of International Relations and Cooperation, Maite Nkoana-Mashabane, established an ongoing Strategic Dialogue to foster cooperation in areas of mutual concern. South Africa is an active participant in U.S. Africa Command's primary maritime security engagement initiative, Africa Partnership Station (APS).

Integrated Interagency Team

U.S. Africa Command is part of a diverse interagency team with abundant talents, expertise, and capabilities. An integrated interagency effort requires understanding the institutional perspectives of each agency, as well as each agency's statutory responsibilities and authorities.

U.S. Africa Command's contribution to an integrated approach starts in our Stuttgart headquarters, where the Deputy to the Commander for Civil-Military Activities is a senior U.S. diplomat. In addition, a U.S. Agency for International Development (USAID) Senior Development Advisor helps ensure that our military programs and activities support and complement our government's development programs and activities. We have made progress in creating a collaborative interagency environment in U.S. Africa Command. While we welcome the skills and capabilities our interagency partners bring to our mission, we are still working toward fulfilling our vision of an integrated whole-of-government approach to the challenges in Africa. We would benefit from increased interagency support from other USG agencies and departments.

U.S. Africa Command also works closely with embassy country teams to ensure that our programs and activities support and complement their Mission Strategic and Resource Plans. To enhance this cooperative interagency effort, our annual Theater Security Cooperation Conference brings together over 300 stakeholders from U.S. embassies in Africa, USG agencies, and our component commands to ensure our programs and activities are synchronized and integrated with those of the other government agencies.

Programs and Activities

U.S. Africa Command is committed to the delivery and sustainment of effective security cooperation programs that build our partners' security capacity. At our partners' request, we design and develop programs, activities, and exercises that support their four common defense goals. When we complete an activity, we stay engaged with our partners to foster long-term gains and successes. Our intent is to achieve a persistent and sustained level of engagement.

Our Army, Navy, Air Force, Marine and Special Operations forces as well as CJTF-HOA set the conditions for the success of our security cooperation programs and activities on the continent.[2] They perform detailed planning, provide essential command and control, establish and sustain relations with our partners, and develop processes to provide timely operational assessments.

Partner security capacity building programs focus on three overlapping capacity building functions:

- Building operational capacity;
- Building institutional capacity; and,
- Developing human capital.

Building Operational Capacity

Building Operational Capacity is about more than the number of troops and pieces of equipment. It is about aligning the right military capabilities—ground, maritime, and air—against a partner's unique mission requirements, as well as developing the necessary operational enablers that foster the independent sustainment of those capabilities. We work closely with our partners to align the right military capabilities against threat and mission requirements, as well as develop the necessary operational enablers. Examples of programs and activities that help our African partners build operational capacity, as mentioned earlier, include OEF-TS, exercise FLINTLOCK, PREACT, and the GPOI/ACOTA programs

Beyond these important programs and activities, the National Guard State Partnership Program, APS, the African Maritime Law Enforcement Partnership, exercises AFRICA ENDEAVOR and NATURAL FIRE, Medical Capabilities and Readiness Exercises, and Africa Deployment Assistance Partnership Team all contribute to partner operational capacity.

The State Partnership Program connects a U.S. state's National Guard to an African nation for military training and relationship building. This program builds long-term relationships, promotes access, enhances African military professionalism and capabilities, strengthens interoperability, and enables healthy civil-military relations. Our current National Guard partnerships are: Botswana and North Carolina, Ghana and North Dakota, Liberia and Michigan, Morocco and Utah, Nigeria and California, Senegal and Vermont, South Africa and New York, and Tunisia and Wyoming. Working closely with the National Guard Bureau, the Command will seek to expand this highly effective program.

Africa Partnership Station (APS) is U.S. Africa Command's primary maritime security engagement program. APS strengthens our partners' maritime security capacity by focusing on the development of planning skills, maritime domain awareness, response capabilities, and regional integration. In 2010, APS included representatives from nine European allies, 23 African countries, and Brazil. This was more than double the number of partners participating in the planning and execution in previous years.

African Maritime Law Enforcement Partnership (AMLEP) is a maritime partner program conducted with the United States Coast Guard that contributes to operational as well as institutional capacity. AMLEP develops partner capacity in maritime governance and maritime security by engaging partner nations at both the national interagency level and tactical enforcement level. AMLEP operations result in the apprehension of vessels engaged in illegal operations and enable U.S. diplomatic efforts that foster good governance.

AFRICA ENDEAVOR is our premier communications exercise designed to encourage interoperability, information exchange, and regional cooperation among African nations so they can coordinate with one another during natural disasters and emergencies. Last year in Ghana, communications experts from more than 30 African nations participated in the continent's largest communications interoperability exercise. A highlight was the direct satellite link and associated command and control tactics, techniques, and procedures established between the African Union's Peace Support Center in Addis Ababa and the exercise main body in Accra, Ghana.

Joint and multinational exercise NATURAL FIRE seeks to improve interoperability and build partner capacity to respond to complex humanitarian emergencies, specifically planning for possible pandemic influenza outbreaks. The fiscal year 2010 (FY10) exercise included more than 500 U.S. soldiers training alongside more than 600 troops from Burundi, Kenya, Rwanda, Tanzania, and Uganda. The FY11 exercise will be held for the first time in Tanzania and is scheduled for July and August 2011.

Medical Capabilities and Readiness Exercises, referred to as MEDFLAG, provide medical training to the host nation's medical personnel as well as offering clinics for residents in the local area. A September 2010 MEDFLAG bi-lateral exercise in Kinshasa, DRC, saw 90 service members from U.S. Africa Command and its components train more than 200 members of the Armed Forces of the Democratic Republic of the Congo in combat lifesaving skills, preventative medicine, and mass casualty procedures. As part of the exercise, doctors from the DRC worked side-by-side with their American counterparts to treat more than 1800 Congolese in a three-day clinic.

Africa Deployment Assistance Partnership Team assists African partners to build logistics capacity by training African partner non-commissioned officers how to deploy their equipment in support of peacekeeping operations. Of the eight partner nations who completed the training in 2009, the Ugandans were the first to get their 23 non-commissioned officers certified to plan, palletize, and load cargo on peacekeeping missions.

Building Institutional Capacity

To support the building of institutional capacity, we focus on developing and sustaining a government's ability to program and allocate security resources, establish national command and control, provide civilian oversight of military forces, and develop military and security recruiting, training, and sustaining programs and policies. These functions help to ensure the readiness and independent sustainability of our partners' military forces. An underlying premise of our building institutional capacity efforts is that military forces must be subordinate to civil authority and accepted as legitimate members of a civil society based on the rule of law.

We have learned from Operation ONWARD LIBERTY in Liberia that rebuilding a state's security institutions requires more than military means. Institutional capacity building is an inherently interagency endeavor. Success in this mission requires a long-term commitment by numerous USG agencies and sufficient resources in all departments to ensure success. ONWARD LIBERTY illustrates how we can achieve positive effects that enhance the capability of our partners and support U.S. interests disproportionate to our modest investment in resources. With a small investment, the USG is helping Liberia transform their Ministry of Defense, Armed Forces, and Coast Guard. Liberia's progress will contribute to regional stability and provide a model for other African states transitioning from conflicts. Moreover, these lessons will inform how we approach Security Force Assistance in other areas of Africa, such as Southern Sudan.

Other programs and activities that are building institutional capacity include Partnership for Integrated Logistics Operations and Tactics, The Pandemic Response Program, Military Intelligence Security Cooperation, and special staff assistance and mentoring programs.

Partnership for Integrated Logistics Operations and Tactics is an operational-level seminar jointly designed and funded in partnership with the Canadian Ministry of Defense, Pearson Peacekeeping Centre and the GPOI program. This program builds long-term operational logistics planning capacity within the AU Standby Force, and promotes interoperability with the U.S. military.

The Pandemic Response Program focuses on strengthening our African partners' military capacities to plan and respond to potential pandemics. Our efforts are reinforced with three years of funding from USAID, which cooperates with the International Federation of the Red Cross and Red Crescent Societies and other partner organizations in African countries.

Military Intelligence Security Cooperation develops not only institutional capacity, but also human capital by enhancing partner country military and security service professionalism and the development of military intelligence organizations that respond to civilian authority and respect the rule of law. Military intelligence operational capacity building is designed to support the execution of full-spectrum operations, including counter terrorism, intelligence support to peacekeeping operations, and intelligence support to humanitarian assistance and disaster relief.

To support the development of institutional capacity, we also conduct special staff programs. Through mentoring and exchanges, our inspector general, chaplain, legal counsel, surgeon, public affairs, and other special staff elements work closely with partners to improve readiness and professionalism. Additionally, our African partners recognize that the backbone of a professional military is an effective non-commissioned officer (NCO) corps. Depending on the need, we help develop doctrine, training plans, and facility plans while also engaging with partner leadership in developing their NCO corps.

Developing Human Capital

To develop human capital, we focus on developing the professional attributes and values that complement capacity and institution building efforts and enhance the standing of the military among members of civil society. We encourage our partners to develop the capacity to take care of their military forces and their families, which increases readiness. Each positive engagement and activity alongside our forces helps our partners develop the professional attributes and values essential for an effective military.

The African Leader Development Initiative is a U.S. Africa Command program that assists in the strengthening of our African partner's warrant officer and noncommissioned officer corps. U.S. Africa Command and our components accomplish this by providing warrant and noncommissioned officer academies with curriculum development assistance, instructors, guest speakers, and familiarization visits for our African partners to our Services' premier NCO academies. Additionally, 2010 saw the first annual African Defense Joint Warrant Officer/Sergeant Major Symposium convene in Washington, D.C. This symposium brought together senior warrant and noncommissioned officers from seventeen African countries to collaborate on civil-military relations, regional security issues, peace support operations, Human Immunodeficiency Virus prevention, sexual and gender based violence prevention, and initiatives for women in the armed forces. This is an example of the outstanding support we receive from our Africa Center for Strategic Studies (ACSS), which co-hosted the 2010 symposium as well as many other events. U.S. Africa Command relies on the social, cultural, and academic expertise ACSS provides to inform our cooperative efforts with our sub-Saharan African partners.

The Military Intelligence Basic Officer Course-Africa, the Military Intelligence Professionals Course, and the Military Intelligence NCO Course-Africa are U.S. Africa Command-sponsored courses that assist partner nations in further developing fundamental military leadership and intelligence skills. They enhance capacity for intelligence analysis and sharing among nations and to provide an environment designed to improve collaboration within the region's military intelligence community.

International Military Education and Training (IMET) is a long-standing and well-regarded program funded by the Department of State. IMET provides military course education and training for foreign military and select civilian personnel. Professionalizing militaries and reinforcing the democratic value of elected civilian authority are among the benefits of the Department of State-led IMET and Expanded-IMET programs. These comprise the most widely used military assistance programs in U.S. Africa Command's area of responsibility as almost every country in the AOR receives IMET. Sending African officers to our mid-level and senior staff colleges provides these officers an opportunity to not only learn about the United States and develop long-lasting professional relationships with our very best officers, but to assume greater responsibilities in meeting their security challenges upon return to their own country. Presently, 30 West and Central African flag and general officers have benefited from the IMET program. A notable IMET participant from Uganda is Major General Nathan Mugisha—the commander of the African Union's peacekeeping mission Somalia; a U.S. Army War College graduate.

Resourcing for the Future

U.S. Africa Command maintains a long-term commitment to our partners to ensure that stability becomes self-sustaining on the continent. Simultaneously, there are potential crises requiring prompt, decisive action, and U.S. Africa Command must be ready for these contingency operations.

As we assess our capacity to execute our operations, programs and activities, we continuously evaluate our progress in four key categories:

- Our ability to foster sustained engagement;
- Our ability to expand our operational reach;
- Our ability to respond rapidly to crisis; and,
- Our ability to take care of our people.

Fostering Sustained Engagement

The 2010 QDR identified building security capacity of partner states as one of six key DOD mission areas. U.S. Africa Command is developing, executing, and refining a range of programs and activities that help African states build capable, professional militaries.

Resourcing Sustained Engagement: Authorities

As the Chairman of the Joint Chiefs of Staff and the Secretary of Defense highlighted on numerous occasions, the strict compartmentalization of engagement funding sources can impede unity of effort and progress. Authorities are often inflexible and processes too cumbersome for today's security challenges. For example, the at-sea training platform APS involved the cobbling together of over a dozen different funding sources from Title 10 to resource its first three years of this important program. This made establishing and administering this important program a challenge. We have had tremendous success with APS, and today the Department of State provides dedicated Title 22 funding to implement the training aspect of this program.

The Global Security Contingency Fund proposed by the President offers a new approach to funding important assistance activities in an effective manner—pooling the resources and expertise of the Department of State and DOD. The Administration has requested $50 million in the State Department budget for this fund, and the Department of Defense has committed to contribute significant funding as well. This approach would create a new business model that we believe will lead to collaborative programs to respond to emergent challenges and opportunities.

Resourcing Sustained Engagement: Forces

Maintaining a predictable pattern of available operational forces for sustained engagement activity is critical to the success of our efforts on the continent and an ongoing challenge. U.S. Africa Command's components and subordinate commands are the primary implementers of our programs and activities on the continent. In an environment of competing global demands for forces, we rely on the Global Force Management process for the necessary support of our engagements in Africa.

Funding for the Foreign Military Financing Program

Foreign Military Financing (FMF) is fundamental to our strategy of preventative rather than reactive response and represents an investment in critical relationships. Inadequate funding or inconsistent year-to-year distributions can compromise our efforts, turn our partners towards other sources, and inhibit peacekeeping and counterterrorism operations. The demand for FMF funds will always exceed resource availability. Therefore, U.S. Africa Command carefully prioritizes its FMF recommendations to the Department of State using a rigorous analytical process that considers national interests, DOD guidance, country team recommendations, host nation desires, and the host nation's capacity to absorb and effectively employ military equipment. For example, Senegal uses this program to maintain the fleet of U.S. military trucks they acquired for peacekeeping operations using the Excess Defense Article program.

Planning and Assessing Our Programs and Activities

U.S. Africa Command is a judicious steward of the resources provided to us by the American people to accomplish our theater objectives. With the fiscal challenges we face at home and the competing global demands on our military forces, we are dedicated to helping build partner nation capability at the lowest cost possible. In many instances, we achieve positive effects that enhance the capability of our partners and support U.S. interests with a modest investment in resources. We do this through a detailed planning process that links our programs and activities to our theater objectives and carefully justifies resources.

While it is difficult to attribute improvements in our partners' security capacity to individual activities or engagements, the evidence indicates that we are moving in the right direction with regard to our direct engagements. In CJTF-HOA, for example, we measure our effectiveness in several ways: socio-cultural research teams, informal and formal feedback from partner nations, and through surveys. We take a broad view of our programs and activities with a robust annual command assessment process, which incorporates indicators from a range of quantitative and qualitative sources and helps guide our planning for future programs and activities.

Funding for Exercises

A key component of our capacity building is our Joint and Combined Exercise program. This program is dependent upon funding from the Combatant Commander's Exercise and Engagement and Training Transformation (CE2T2) Program. These exercises improve not only the readiness of our African partners, but also increase the capabilities of U.S. forces. As our exercise program expands to meet the readiness needs of U.S. forces and partner militaries, U.S. Africa Command will place increasing demands on limited CE2T2 funds. We seek Congress' continued support of the DOD's request for the Combatant Commander's Exercise, Engagement, and Training Transformation Program.

Though we face challenges in the previously mentioned areas, we are also entering into a new era of opportunities.

Quadrennial Diplomacy and Development Review

The Department of State's recent Quadrennial Diplomacy and Development Review (QDDR) is a laudable milestone on the pathway to improving interagency cooperation and collaboration. Following in the footsteps of the Department of Defense's QDR, the QDDR made a sweeping assessment of how the Department of State and USAID could advance our national interests and effectively partner with the U.S. military in support of these interests. The QDDR's elevation of conflict prevention and response to a core civilian mission is notable, as is the commitment to developing an integrated capability to reform security and justice sectors in fragile states. We look forward to working with our interagency partners at the Department of State and across the government as they implement the QDDR's recommendations.

African Cooperation Authority

U.S. Africa Command makes maximum use of Traditional Combatant Commander Activities authorities for military-to-military and other activities. Section 1050a, a new revision to Title 10, United States Code, will be an important additional authority enabling engagements with civilian elements of Ministries of Defense, the AU, and other security organizations. This authority, provided through the Ike Skelton National Defense Authorization Act for Fiscal Year 2011, will open doors to broader cooperation and benefit both the United States and our African partners. We welcome this valuable addition and thank Congress for its support in creating this authority.

Enhancing Operational Reach and Ensuring Rapid Response to Crisis

U.S. Africa Command works to enhance our operational reach across the Command's AOR. This supports our ability to respond to crises promptly and our long-term efforts for security as well. U.S. forces are often employed in austere environments with limited to non-existent infrastructure, port, and logistics networks. Improvements in these areas enhance our ability to respond to emergencies and threats and to meet the needs of sustained engagement.

Theater Infrastructure and Posture Requirements

U.S. Africa Command evaluates and refines its access needs on a regular basis, in close cooperation with the Department of State. This effort centers on gaining and maintaining the access and freedom of movement necessary to conduct day-to-day security cooperation activities and crisis response operations. We work with our components, the Office of the Secretary of Defense, and the Department of State to update the network of Cooperative Security Locations and supporting agreements required to enable the Command to carry out its activities. As U.S. Africa Command's sole Forward Operating Site on the continent, Camp Lemonnier, Djibouti, serves as a critical operational and sustainment facility (a hub for lift, logistics, intelligence, and Command, Control, Communications and Computers (C4)) in support of DOD activities in the region. It is critical for activities and operations across AOR boundaries in the Red Sea, the Gulf of Aden, and the Arabian Peninsula.

Camp Lemonnier's proximity to Djibouti's air and seaport make it ideal for supporting U.S. Africa Command operations throughout the region. Of equal importance is the Camp's ability to support DOD's global transportation infrastructure network as a key node. Camp Lemonnier, Djibouti, is essential to U.S. security interests in East Africa and the Gulf of Aden. The FY12 military construction request contains a number of important projects that will allow us to continue to enhance the capabilities of this facility. We appreciate your attention to these requests.

Ensuring Rapid Response to Crisis

U.S. Africa Command conducts prudent operational planning for a range of possible humanitarian and security contingencies beyond fostering sustained engagement and working to expand our operational reach. This requires the staff to continually assess potential force requirements and the lift, logistics, C4, and intelligence, surveillance, and reconnaissance enablers required to support a range of contingencies. We share our assessments regularly with Joint Staff and the Office of Secretary of Defense.

Taking Care of Our People

U.S. Africa Command team members and their families serve in Stuttgart, Alconbury (Molesworth), and across Africa. Quality of Life (QoL) Programs affirm our commitment to our team members and their families wherever they serve. This Command is committed to providing a strong, supportive environment fostering growth and excellence, while providing the highest quality of resources and services to our U.S. Africa Command family consistent with their commitment and dedication to serve the Nation. To that end, we created a QoL office to manage and oversee activities at the Stuttgart headquarters, Alconbury, and on the African continent.

In order to understand the needs of our families stationed throughout the continent, U.S. Africa Command hosts the annual 'Africa Command Families on the African Continent' working group. This annual forum addresses emerging QoL issues and provides the foundation for our QoL Action Plan. This forum recently identified two issues that family members highlight on a regular basis. First, family members have requested changes that would allow expectant mothers to elect an alternate destination for obstetric care, providing an option to be with their families instead of alone at the nearest Military Treatment Facility. The second issue is the provision of student travel benefits for dependents attending accredited overseas colleges or universities off the continent in locations such as Europe. We appreciate Congress' attention to these two important issues.

To further assist our team members and their families in solving problems resulting from deployment, reunions, and other family changes, U.S. Africa Command implemented the Military and Family Life Consultant Program to support the Command headquarters, remote locations and the African continent. We currently have three consultants supporting the Stuttgart and Alconbury communities, as well as CJTF-HOA in Djibouti. We continue to focus our efforts on our members and their families, both on and off the African continent, to ensure their quality of life remains a priority and is funded properly.

Conclusion

U.S. Africa Command is protecting American lives and promoting our national interests today by helping our African partners assume a continually increasing role addressing their own security concerns. Africa's long-term growth, stability, and economic prosperity is largely dependent on our partners' ability to develop capable and professional militaries that are subordinate to civilian authority, respect human rights, and adhere to the rule of law. There are no short cuts to growth and stability in Africa; we must shape our efforts with an understanding of the continent and patiently sustain our efforts over time.

A prosperous and stable Africa is strategically important to the United States. An Africa that can generate and sustain broad based development will contribute to global economic growth and vitality—a long-standing American interest. Prosperity and stability in Africa will ensure that it does not become a haven for those who wish to do harm to our citizens and our interests—both in the homeland and abroad.

In the coming year, we will continue to support African leadership in addressing shared security challenges, take advantage of opportunities, deepen our strategic partnerships—regionally, bilaterally, and with the AU—and refine our focus in our security cooperation efforts, while also ensuring the military readiness and operational capability of U.S. Africa Command.

I am grateful for the outstanding support U.S. Africa Command has received from the Congress. The continued support for our uniformed men and women, our DOD civilians and the other USG departments and agencies assigned to the command will allow their tremendous work to continue. I am proud to serve on the U.S. Africa Command team with these dedicated Americans.

End Notes

[1] Egypt is in U.S. Central Command's AOR. The United States does not recognize the Sahrawi Arab Democratic Republic (Western Sahara); however, it is recognized by the African Union.

[2] U.S. Africa Command has four component commands, one sub-unified command, and the Combined Joint Task Force — Horn of Africa. The four component commands are U.S. Army Africa (USARAF), U.S. Naval Force, Africa (NAVAF), and U.S. Air Forces (AFAFRICA/17AF). U.S. Special Operations Command, Africa (SOCAFRICA) is the sub-unified command.

In: Combatant Commands in the United States Armed Forces ISBN: 978-1-61942-924-6
Editors: Quincy N. Butler and Harris A. Simmons

Chapter 3

STATEMENT OF GENERAL JAMES N. MATTIS, UNITED STATES MARINE CORPS COMMANDER UNITED STATES CENTRAL COMMAND BEFORE THE SENATE ARMED SERVICES COMMITTEE ON THE POSTURE OF UNITED STATES CENTRAL COMMAND*

1 March 2011

I. INTRODUCTION

A Command at War

U.S. Central Command (CENTCOM) oversees operations alongside our allies, partners, and friends in a critically important region of the world. CENTCOM is engaged throughout the greater Middle East and South Central Asia across the full spectrum of warfare, standing against violent aggression and the tyranny of militant extremists, while contributing to the broader conditions for peace, stability, and prosperity.

Recognizing our Troops, Civilians, and Partners

Our troops and their families carry the brunt of physical and emotional burdens in this tenth year of war. Today, over 200,000 American troops and tens of thousands of civilians are deployed to the CENTCOM AOR. These men and women – all volunteers, no less – defend our freedoms with great courage in the face of a murderous enemy on harsh terrain. And our troops stand together with tens of thousands of our international partners, conducting

* These remarks were delivered as testimony given on 1 March 2011. General James N. Mattis, U.S. Marine Corps Commander, U.S. Central Command before the Senate Armed Services Committee on the Posture of U.S. Central Command.

coalition operations from the waters off Somalia to the mountains of Afghanistan, where the largest war-fighting coalition in recent history is engaged.

Operating in a Dynamic Region

The CENTCOM AOR is more dynamic than I have seen it since first serving there in 1979. Across our theater, we are required to maintain a degree of military flexibility such as we have seldom seen before. At the same time, given the financial realities in Washington, we require ourselves to exercise the utmost degree of stewardship over every penny we spend. To operate in this context successfully, we seek to build strong military-to-military relationships with our partners, recognizing that CENTCOM's actions represent a tangible signal of America's continued, long-term commitment to the security and prosperity of this area.

Throughout the region, we see institutions of government responding to the aspirations of youthful populations. As the people in the region have made their voices heard, regional militaries have so far demonstrated their professionalism, exercising a capability that did not arise by accident or overnight. The strong security relationship between the U.S. and our partners is decades in the making and has helped them become the professional forces they are today – and in the process made our forces better as well. While we seek to understand the unique circumstances that our partners confront, CENTCOM remains committed to supporting the efforts of our military counterparts and to strengthening the security partnerships that have proven critical during this period of political unrest. We do this first by listening, learning, and understanding, and continue by engaging with our partners based on mutual respect and shared interests.

Our Mission

Overall, amidst these conditions, we remain committed to carry out our mission:

> With our national and international partners, CENTCOM promotes security cooperation among nations; responds to crises; deters or defeats state and non-state aggression; supports development and, when necessary, reconstruction in order to establish the conditions for regional security, stability, and prosperity.

Snapshot of Operations

Our main effort is Afghanistan – and progress there is indisputable, even if some of our success is fragile and reversible. We and our North Atlantic Treaty Organization (NATO) and coalition partners are conducting a comprehensive yet focused counterinsurgency campaign to ensure Afghanistan does not once again become a sanctuary for transnational extremists. Our forces are part of a 49- nation international coalition, led by the NATO International Security Assistance Force (ISAF), and united behind President Karzai's goal of transitioning the lead of security tasks from the international community to Afghan security forces by the end of 2014. In full partnership with the Afghan government, we are inflicting unprecedented damage on al-Qaeda (AQ) and associated extremist groups – a reality recently affirmed by President Obama's Afghanistan-Pakistan Annual Review. Moreover, we confound our enemies by demonstrating our unambiguous commitment to our long-term strategic partnership with Afghanistan. Meanwhile, in Pakistan, we continue supporting Pakistan's military efforts against extremists operating from and threatening that country and Afghanistan, while contributing to the broader U.S. goal of growing our strategic partnership with Islamabad. The recent U.S.-Pakistan Strategic Dialogue and a number of development assistance programs sponsored by the Department of State are good examples of how the U.S. is attempting to build trust with the Pakistani people and government.

In Iraq, following seven years of hard fought gains, we are drawing down our troops as we transition full security responsibilities to our Iraqi partners. The enemy in Iraq is capable of dramatic attacks but has proven unable to muster a significant threat to the Iraqi government. In coordination with the U.S. Department of State, CENTCOM is standing up the Office of Security Cooperation-Iraq to conduct sustained security assistance and cooperation activities with the government of Iraq. We are planning an organization manned and positioned to support the long-term U.S. objectives in Iraq as determined by the Iraqi and American governments, in order to best advance our civilian-led relationship for the future.

In the broader CENTCOM region, our forces are conducting a theater-wide campaign alongside our partners in pursuit of AQ and its extremist allies. Meanwhile, we remain continuously poised and postured to respond to crises and to conduct contingency operations, while continuing to forge partnerships in the region and increase the security capacity of our partners. We continue to rely on our capable and flexible amphibious forces. For example, over a 36 hour period last September, the 15th Marine Expeditionary Unit delivered aid to the flood-ravaged people of Pakistan, provided close air support from the skies over Afghanistan, and rescued pirated crews in the Gulf of Aden. Three months later, two-thirds of our Marines Expeditionary Unit deployed to Afghanistan on three day's notice.

I. Overview of the CENTCOM AOR

A. Nature of the AOR

The CENTCOM AOR is comprised of 20 countries spanning over four million square miles in three diverse sub-regions from Egypt and the Levant, to the Arabian Peninsula (including the Gulf nations), and Central and South Asia. These regions are home to a half-billion people practicing all of the world's major religions and speaking more than 18 major languages. Several countries with economic challenges have burgeoning populations – 184 million people in Pakistan, 80 million in Egypt, and 77 million in Iran. In 12 of the 20 countries in the region, 30 or more percent of the population is between the ages of 15 and 24 (at 39 percent, Yemen ranks at the top in this category). In most of those countries, another 30 percent of the overall population is under 15. This youth bulge represents tomorrow's future leadership and the region's greatest challenge in terms of education, employment and expectations.

The CENTCOM AOR is a region of rich history, distinct culture, and great potential, encompassing the proud traditions of a wide variety of ethnic groups, including: Arab, Azeri,

Baluch, Gilaki, Hazara, Kurd, Lur, Mazandarani, Qashqai, Pashtun, Persian, Talysh, Turkmen, and Uzbek, among others. The AOR contains more than half of the world's proven oil reserves and nearly half of its natural gas. As a result, the region contains some of the world's busiest trading routes linking Europe, Africa, and East Asia to the Gulf. This trade is essential to continued global economic prosperity and growth. The region's trading routes contain three of the world's major maritime choke points, including the Strait of Hormuz, the Suez Canal, and the Bab al Mandeb Strait joining the Red Sea to the Gulf of Aden. But while the region contains abundant energy resources, supplies of water and the availability of arable land are limited and increasingly scarce.

B. External Influences on the CENTCOM AOR

The region retains its historical tradition as a social, economic, and cultural crossroads, attracting nations and non-state actors seeking to advance their interests and influence regional events. Among a host of external influences on the CENTCOM AOR, the most significant include:

- Middle East Peace : Lack of progress in achieving comprehensive Middle East peace affects U.S. and CENTCOM security interests in the region. It is one of many issues that is exploited by our adversaries in the region and is used as a recruiting tool for extremist groups. The lack of progress also creates friction with regional partners and creates political challenges for advancing our interests by marginalizing moderate voices in the region. As Secretary Gates noted in July 2010, "the lack of progress in the peace process has provided political ammunition to our adversaries in the Middle East and in the region, and...progress in this arena will enable us not only to perhaps get others to support the peace process, but also support us in our efforts to try and impose effective sanctions against Iran." In December 2010, Secretary Clinton observed "the conflict between Israel and Palestine and between Israel and its Arab neighbors is a source of tension and an obstacle to prosperity and opportunity for all of the people in the region." By contrast, substantive progress on Middle East peace would improve CENTCOM's opportunities to work with our regional partners and support multilateral security efforts. Speaking about the need for Middle East peace at the Manama Dialogue in December 2010, King Abdullah of Jordan observed "Our region will not enjoy security and stability unless we solve the Palestinian-Israeli conflict, and Arabs and Israelis find peace. The stakes are high. As a solution continues to elude us, faith in negotiations, as the only path to peace and justice, is eroding. And if hope is killed, radical forces will prevail. The region will sink into more vicious warfare and instability...threatening security far beyond the borders of the Middle East."
- *Bordering Powers.* China, Russia, Turkey, and India – each of which lie outside but border the CENTCOM region – represent four great gravitational forces influencing various countries in the AOR. China pursues its many energy-related interests throughout the region, extending influence from its traditional partnership with Pakistan, to a $3.5B investment in Afghanistan's Aynak Copper Mine, to building

pipelines for oil and gas from Kazakhstan to Turkmenistan. Chinese activities in the region may begin to compete with the regional interests of Russia, which maintains a network of security, economic, and social ties with Central Asian nations and beyond. India's influence impacts the strategic calculations of Pakistan and, to some extent, virtually every other country in the CENTCOM AOR. Turkey increasingly asserts its interests in the region in keeping with its emergence as a considerable force within the international community. All four of these nations have unique relations with Iran, affecting the international approach to the Iranian situation. We remain attentive to these dynamics as we seek to ensure that we work effectively across U.S. government and Combatant Command seams to improve our unity of effort.

- Somalia. State failure in Somalia has enabled extremist and criminal elements to proliferate and spread northward into the Horn of Africa and Yemen and other areas of the CENTCOM AOR. At the same time, widespread poverty in Somalia creates incentives for young men to pursue the lucrative enterprise of piracy. Additionally, lack of governance permits extremists to freely migrate to Yemen, providing opportunities to al-Qaeda in the Arabian Peninsula (AQAP). In the past year, the Somalia-based terrorist group al-Shabaab successfully maintained control of most of southern Somalia and radicalized factions of this group have sought alignment with AQ in the Arabian Peninsula and in Pakistan.

C. U.S. Interests in the Region

Given the centrality and volatility of the CENTCOM AOR, the U.S. and nations around the world retain significant interests in the region. Among others, significant U.S. interests in the region include:

- Security of U.S. citizens and the U.S. homeland
- Regional stability
- Promotion of effective and legitimate governance, human rights, the rule of law, and sustained economic growth and opportunity, and
- Free flow of commerce and trade within the region, through strategic maritime chokepoints, and via land-based trade routes to international markets

D. Threats to U.S. Interests in the Region

Violence, instability, and underdevelopment represent the primary threats to U.S. interests in the region. Some areas face uneven or even dismal economic development, often coupled with endemic corruption. Social and economic friction have led to or exacerbated a number of deep-rooted and long-standing disputes over territory, resources, and power, many of which remain unresolved due to a lack of adequate security arrangements on the local or national level. Some areas will face increasing competition for food, water, mineral deposits, oil, and other natural resources. The region is also defined by tensions and sectarian rivalries between many ethnic, tribal, and religious groups. Such conditions create the potential for broader violence, particularly in the absence of effective governance and indigenous security forces, ultimately giving rise to violent extremist organizations that have attacked us and our friends. We have seen the dangers present within a security vacuum, where institutions fail to facilitate mediation, partnership-building, and open dialogue between feuding groups, or to put down violent extremists.

E. Connecting Our Strategic Challenges

The challenges of the CENTCOM AOR are inextricably linked and mutually reinforcing – and thus cannot be treated separately. We have seen a symbiosis, for example, between extremist groups and other factions that, in aggregate, tend to strengthen each other and which, if left unchecked, tend to threaten wider areas of territory and the stability of civilian governments. Areas in the CENTCOM region, especially those with a rapidly expanding population of youth, are left vulnerable to (and often become the victim of) a worsening spiral of conditions, whereby young people forego meager, but legitimate opportunities for employment and turn, instead, to a range of criminal activities, including piracy, arms smuggling, human trafficking, and narcotics – fueling violent extremist organizations bent on destroying the lives of innocent people. State and non-state actors operating with malign intent can readily exploit such conditions, with the most dangerous scenarios involving a mix of insufficient governance, weapons proliferation – especially Weapons of Mass Destruction (WMD) – the influence of hostile states, and the free flow of extremist elements across national borders as well as free range in cyberspace. In some cases, disenchantment with globalization's efforts coupled with a desire to belong to a movement with a clarion call of purpose can provide the excitement for young men (and increasingly women) to take on a violent role in an extremist organization.

II. Principal Tasks

In light of these many challenges, we continuously assess our strategic and operational approaches in order to achieve our desired national interests of security, stability, and prosperity in the CENTCOM AOR. CENTCOM is focused on the following tasks:

- Supporting the Mission in Afghanistan
- Partnering with Pakistan
- Countering the Destabilizing Activities of Iran
- Enabling Transition in Iraq
- Strengthening Partnerships in Central Asia
- Building Partner Capacity and Pursuing Cooperative Activities
- Disrupting Violent Extremist Organizations
- Combating Weapons of Mass Destruction
- Countering Piracy

Supporting the Mission in Afghanistan

Instability in Afghanistan and Pakistan

Afghanistan and Pakistan are inextricably linked, connected by a porous border region historically providing free movement and safe haven to groups traversing the Durand Line. The senior leadership of AQ and associated extremists groups – groups that are intent on carrying out attacks on innocent civilians worldwide – plan, prepare, and direct operations from this region, making it of critical interest to the security of the U.S. and our allies. Currently AQ in the border region is under the most intense pressure they have experienced since 2001.

A Clear Objective and a Sound Strategy

With our NATO and coalition partners, we are working to achieve our core goal of preventing Afghanistan from once again becoming a sanctuary for al-Qaeda and associated transnational extremist groups. President Obama's Afghanistan-Pakistan Annual Review affirmed the core elements of our strategy in Afghanistan, the first imperative of which is to improve the overall security environment and to reduce violence levels in Afghanistan. After regaining the initiative from the enemy, our forces act as a bulwark behind which the Afghan National Security Forces (ANSF) and the roots of Afghan governance can grow.

Aiming Toward a Common Strategic Vision

Our military objectives and strategy in Afghanistan support the developing strategic vision between the political leadership of the U.S. and Afghanistan, as reflected in Vice President Biden's comments alongside President Karzai in January: "It is not our intention to govern or to nation-build. As President Karzai often points out, this is the responsibility of the Afghan people, and they are fully capable of it. We stand ready to help you in that effort. And we will continue to stand ready to help you in that effort after 2014." Success in Afghanistan is an Afghan security force able to protect the people with a government that meets the needs of the people and prevents safe haven for international terrorists.

The Campaign Plan

We have increased efforts in virtually every facet of the comprehensive yet focused civil-military campaign in Afghanistan. As one part of that effort, we have executed an unprecedented pace of counterterrorist operations to capture or kill insurgents using enhanced intelligence largely enabled by conventional ground forces. Our efforts range from major combat operations (in Helmand and elsewhere, for example), special mission unit operations allowing no safe haven to the enemy, and concurrent bottom-up and top-down initiatives (exemplified by expanding village stability operations).

The breadth of our current operations squelches the enemy's ability to recuperate and threaten the Afghan people. Our forces have partnered with the Afghan security forces to increase their capabilities; to expand border security; to conduct robust detainee operations and rule of law activities; to address and counter corruption by working with the Afghan government to target criminal patronage networks; and to interdict the flow of illegal weapons and narcotics to deny criminals and insurgent groups a critical source of their operational revenue. We are capitalizing on our expanded security footprint in the winter months to retain the initiative, suffocate the enemy, and increase momentum into the start of the traditional fighting season. Ultimately, we are working to create an Afghanistan that is hostile to our enemies and denies them the support of the population, making it untenable for insurgents to return from their winter safe havens. This is the essence of counterinsurgency operations. While we will face tough fighting this spring, the enemy's situation continues to worsen day-by-day.

The Right Inputs

The overall international effort in Afghanistan has transformed from an economy of force mission 3 years ago to a focused and reinforced civil-military counterinsurgency campaign, largely assuming its full strength in September of 2010. U.S., Coalition, and partner nations have worked hard to apply the right mix of organizations, approach, and resources in Afghanistan. Last year at this time, we had less than 270,000 American, coalition and Afghan forces on the ground in Afghanistan. This year, we have more than 370,000 total security forces (American, coalition and Afghan) in the fight, and 109,000 Afghan security forces are projected to be added by this time next year. Beyond the additional organizations put in place on the ground in Afghanistan, the Pentagon's Joint Staff Pakistan Afghanistan Coordination Cell and CENTCOM's Afghanistan Pakistan Center of Excellence are better organizing our resources at home and providing mission-critical reach-back support to deployed forces. The CENTCOM Center of Excellence will provide the cadre of regional experts for the long haul as we transfer to Afghan lead in 2014 and commit to a long-term partnership with Afghanistan and Pakistan.

Enemy Violence and Coalition Progress

Despite the enemy's efforts to disrupt progress in Afghanistan, we have achieved the major military objectives we set out to accomplish in 2010 and made considerable progress with respect to governance and development. As Secretary Gates noted after his December 2010 trip to Afghanistan: "The bottom line is that in the last 12 months, we have come a long way. Frankly, progress—even in the last few months—has exceeded my expectations." We recognize, however, that progress and violence coexist in this type of war. Our enemies continue to conduct attacks heavily focused on non-combatants and to intimidate the population and maintain relevancy, albeit decreasing, in newly-cleared areas. And enemy-initiated violence is increasingly localized. From November 2010 until 31 January 2011, 57 percent of the violence in Afghanistan has been concentrated in 12 of 401 districts. Notably, the key districts of Maiwand in Kandahar Province and Lashkar Gah in Helmand Province – which are critical to our efforts to link the Helmand and Kandahar security bubbles – are no longer among the top-12 most violent districts. The elevated levels of violence is less a reflection of increased insurgent capability and more the result of increased Afghan and ISAF operations in areas previously considered insurgent strongholds. The enemy is not adapting well to this development. While we make progress, our enemies continue to make grievous mistakes, to include: purposefully killing innocent Afghans; leaders fleeing into Pakistan and leaving subordinates to fight; and killing nearly 5,000 Afghans in the first ten months of 2010 (more than three quarters of all civilian casualties in that period). We highlight the ruthless actions of the enemy, and in recent months Afghan leaders and human rights groups have stepped forward to condemn insurgent-initiated violence.

Road to Transition in 2014

We and our North Atlantic Treaty Organization (NATO) and other Coalition and ANSF partners are improving security for the Afghan population, increasing the size and quality of the ANSF, and supporting efforts to improve governance and development throughout Afghanistan. At last November's NATO Summit in Lisbon, we undercut a key pillar of the Taliban's strategy by affirming the long-term resolve of the U.S. and international community to accomplish the mission in Afghanistan. We are united in support of President Karzai's goal of Afghan forces assuming security responsibilities from the international community by the end of 2014. In partnership with the Afghan government, we are working toward President Obama's goal of beginning a drawdown of U.S. forces from Afghanistan beginning in July of this year at a pace determined by conditions on the ground. The process for identifying, assessing, and transitioning areas of Afghanistan is based on recommendations from the Joint Afghan-NATO *Inteqal* (Transition) Board (JANIB) to the Government of Afghanistan. ISAF is working closely with JANIB as we begin the process of transition and methodically move forward in our campaign.

ANSF Suppor

Most importantly in the security arena, our investment in the ANSF is working and the growth of the force is on track. The remarkable quantity growth of ANSF (rising by an unprecedented 70,000 personnel while facing a determined enemy) is now being matched by quality improvements in the force. NATO Training Mission-Afghanistan is supporting the efforts of the Afghan government to build leaders at all levels, to increase literacy, and to improve capability and training capacity. Combined, these programs increase the quality of the force, ultimately helping to reduce attrition, enhance recruitment, and contribute to sustainability. Meanwhile, we are helping the ANSF to overcome remaining challenges in the recruitment of medical staff and other enablers, as well as increasing the participation of females and recruiting more southern Pashtuns. In league with Admiral Stavridis (Commander, U.S. European Command and Supreme Allied Commander Europe), we are trying to reduce our shortage of trainers.

ANSF in the Lead

The ANSF is increasingly in the lead of operations in many areas of Afghanistan. In southern Afghanistan, the ANSF took the lead in mid-2010 for an operation in Malajat, Kandahar City – with support from ISAF for additional combat power, close air support and other enablers – resulting in the capture or killing of several dozen insurgents and the establishment of a new model for Afghan-led operations. The ANSF also provided well over half of the combat power for the latter phases of Operation Hamkari, clearing the insurgency's most vital safe havens in southern Afghanistan. In northern Afghanistan, Afghan National Army and Police conducted joint operations throughout December 2010 with ISAF forces in northern Balkh Province, and Afghan National Police have demonstrated considerable capacity by capturing insurgents and discovering caches of weapons in U.S.-Afghan partnered operations in Kunduz Province. Additionally, ANSF now leads security efforts in 14 of 15 of Kabul's districts, and have executed coordinated security plans for several events, including the June Consultative Peace Jirga, the July Kabul conference,

August Independence Day events and the January seating of the Parliament all without incident, at odds with the insurgents' claims that it would seek to disrupt them.

Local Security Initiatives

Beyond national level security efforts, the Afghan government has steadily expanded the local security initiatives designed to squeeze extremist elements from their traditional safe havens and cut off their lines of communication. Clearing operations in key terrain districts have shifted operational-level momentum and altered village-level calculus in remote areas. Local elders in dozens of villages throughout Afghanistan have conducted jirgas to assume increased responsibility for their own security, and U.S. and coalition forces have supported the Ministry of Interior's efforts to fortify Afghan villages. The Afghan Local Police (ALP) program represents one of the most promising endeavors to wrest local areas from insurgent influence. The ALP and other Village Stability Operation initiatives work from the bottom-up and the top-down, connecting the support of local communities with the capacity of the central government and coalition partnerships. The Taliban has revealed their concerns that the ALP represents a direct threat to their existence and operational ability. Today, there are a total of 63 ALP sites – 24 of which the Ministry of Interior has site validated – and approximately 4,000 ALP are now assigned. These local efforts buttress security in areas with limited ANSF presence, complementing the progress made elsewhere (and in ALP locations) by conventional ISAF and ANSF. Given the initial success of the ALP program, the Ministry of Interior wants to increase the program beyond the current projected number of 10,000 with our reinforced special operations forces providing oversight and mentoring.

Popular Support

Since 2003, AQ and the Taliban have tried with some success to expand their strength and influence in much of the country. In 2010, coalition and Afghan forces applied additional resources in all aspects of the campaign to change the security landscape in much of the country. As security improves in key areas and we are better able to protect the people, Afghanistan's population has increasingly supported efforts to bring development and basic services to their areas. In recent months, in particular, Afghan security forces have assumed more of the load in the fight, village elders have encouraged young men to join the Afghan police, and insurgents in several areas have begun to put down their weapons and integrate into society. Reintegration efforts are bearing fruit due to the concerted effort of the Afghan government both at the local and national level and the support of coalition forces (aided, of course, by the momentum in our campaign). In terms of reconciliation, the process is led by Afghans, with ISAF partnering with ANSF to set security conditions and dash the enemy's hopes of victory. These are progressive steps toward building irreversible momentum in our overall campaign.

Rule of Law Progress

Unlike our enemies, we continue to support the legitimate efforts of the Afghan government to improve the Rule of Law for Afghanistan's more than 29 million inhabitants. U.S. Forces-Afghanistan's Joint Task Force / Combined Interagency Task Force 435 and our Afghan partners have achieved considerable progress in the last year: transferring detainees to the state-of-the-art detention facility in Parwan; implementing transparent and robust internment processes; strengthening judicial guarantees for detainees; and expanding robust reintegration programs that include literacy and vocational training. Moreover, we have established robust efforts to combat corruption at all levels, even as we implement best practices to reduce the challenge of corruption in contracting and in every aspect of our campaign.

Infrastructure Initiatives

We are also pursuing infrastructure initiatives – for example, building roads, rail, and installing electrical grids and transmission lines – to capitalize on Afghanistan's potential as a Central Asian economic hub. A regional transport network facilitates the creation of private sector jobs and provides additional incentives for reconcilable elements of the insurgency to abandon the fight. Ultimately, such economic development reduces the need for U.S. forces and underpins long-term transition activities and is fundamental to a sound counterinsurgency campaign.

Congressional Support

Congressional leadership continues to play a critical role in enabling our efforts in Afghanistan, including the Afghanistan Security Forces Fund (ASFF), the Commander's Emergency Response Program (CERP), the authorization of an infrastructure program, and the Afghanistan Reintegration Program (ARP). Above all, we rely on the ASFF to enable the eventual full transition of security tasks to a robust, trained ANSF capable of preventing the resurgence of insurgent safe havens in Afghanistan. In terms of the CERP, our Commanders on the ground continually comment that the CERP funds are invaluable in carrying out operations toward our strategic objectives in Afghanistan, undercutting the enemy's information operations and legitimacy. In 2010, CERP funded more than 8,300 projects, including, for example, transportation initiatives to improve freedom of movement throughout Afghanistan; agriculture production across Afghanistan involving the repair and improvement of irrigation canals and wells and providing farmers with higher-quality seeds and fertilizers; education projects such as the services of more than 200 local Afghan education outreach coordinators; and water and sanitation projects to install three high-production groundwater wells that will increase the accessibility of potable water to over 850,000 Afghans in Kandahar City. Apart from CERP, the new Afghanistan infrastructure program enables us to work together with the U.S. State Department to undertake high-priority infrastructure projects to address critical needs for Afghan security, governance, and development. The Afghanistan Infrastructure Fund will be the vehicle for the Defense Department's contribution to this integrated program. To enable our reintegration efforts, we continue to execute the ARP using funds in support for the government of Afghanistan's Peace and Reintegration Program.

Challenges Ahead

Much work remains to achieve our goals in Afghanistan. We face a resilient and determined enemy. The U.S. and the international community are positioned to favorably influence reform and synchronize Rule of Law development to counter corruption within the Afghan government. Despite considerable progress in many areas in 2010, we recognize that there will be hard work ahead as we continue to fight along with our Afghan partners. But, by progressively and steadily executing our sound and validated strategy, I believe we can set the conditions to succeed in Afghanistan.

Partnering with Pakistan

Strategic Partnership

We recognize, of course, that any solution in Afghanistan must address the regional context. CENTCOM supports President Obama's goal of strengthening the U.S.-Pakistan strategic partnership through nascent yet improving military-to-military cooperation with Pakistan. And as Secretary Clinton and other leadership has noted, we must concentrate on the efforts Pakistan is taking. They have made very significant moves for going after the terrorist within their own country.

Over the past year, CENTCOM has strengthened and deepened our security cooperation with Pakistan by supporting our counterparts through CENTCOM's Office of Defense Representative-Pakistan (ODRP). ODRP is focused on assisting Pakistan's counterinsurgency efforts and this past year, led the U.S. interagency effort to provide disaster relief and Humanitarian Assistance to areas affected by the flooding. Additionally, in support of our long-term partnership with Pakistan, the CENTCOM Center of Excellence continues to deploy subject matter experts and provide unique reach-back support to ODRP and Special Operations Command-Pakistan (Forward) in order to deepen analysis and to provide greater interagency fidelity on critical issues.

Threats in Pakistan

The potential for instability in Pakistan and the free movement of extremists in the Afghanistan-Pakistan border region continue to pose a serious threat to regional and global security. Pakistan's tribal areas remain the principal sanctuary for alQaeda and a safe haven for other extremist groups, enabling them to threaten the population and coalition forces in Afghanistan, the people and government in Pakistan, and U.S. and Western interests globally. The Afghanistan-Pakistan region also faces significant humanitarian concerns, including refugees and Internally Displaced Persons (IDPs) from decades of conflict. Additionally, roughly three million Afghan refugees still live in Pakistan, having been displaced by the Russian invasion into Afghanistan thirty years ago.

U.S. Humanitarian Assistance

Last summer's historic flooding in Pakistan was devastating – effectively equivalent in scope to flooding the entire East Coast of the United States. The U.S. responded to the floods by providing historic levels of Humanitarian Assistance. In all, U.S. rotary and fixed wing aircraft transported more than 40,000 displaced persons and delivered more than 26 million pounds of aid supplies to the people of Pakistan. U.S. helicopters flew more than 5,000 flight hours during the relief operation. The U.S. government provided Zodiac boat kits to the Pakistan Military for use in rescue operations, and provided eight 50 meter bridges to replace bridges swept away by the floods.

U.S. Support to Pakistan Military

On the security front, continued U.S. assistance is critical to enabling Pakistan to conduct effective counterinsurgency operations. Our forces carry out important partnership and engagement activities in support of the Pakistan military's improving counterinsurgency capabilities.

As one important example, ODRP supports Pakistan's Frontier Scouts by providing training support and enabling further counterinsurgency operations. U.S. personnel also assist in the procurement of materials and equipment needed to build infrastructure in support of education, power, and food.

Pakistan Operations and Sacrifice

Pakistan's military has made impressive strides in combating militants in the FATA, while dealing with the effects of large-scale flooding that devastated much of the country. Over the last year, the enemy has lost battlespace to the Pakistan military's sustained efforts to move against the enemy strongholds. Pakistan's military has suffered more than 2,500 casualties (enduring more than 500 personnel killed in action and more than 2000 wounded in action) since the start of offensive operations against extremist elements in the KPk and the FATA.

Since June 2009, the Pakistan Military has been involved in nearly continuous operations against militants in the KPk and the FATA. In total, the Pakistan Military has deployed upwards of 140,000 troops along Pakistan's western border with Afghanistan, a significant portion of which were drawn from Pakistan's border with India.

Regional Context

Our efforts to support Pakistan fit well within the broader regional context. We recognize that Pakistan's long-standing tensions with India are an important part of Pakistan's strategic decision-making calculus and military force posture. However, the presence of extremist sanctuaries in Pakistan significantly impacts our progress in Afghanistan, and with the Pakistan military's help we are taking important steps to improve cross-border operations. To address existing challenges along the Afghanistan-Pakistan border, coordination between ISAF, Afghan security forces, and the Pakistan Military continues to improve, especially in the area of Intelligence, Surveillance and Reconnaissance (ISR). In Regional Command East, we are planning coordinated operations with the Pakistan Military. The Pakistan Military recently began clearing insurgent safe havens in Mohmand Agency across the border from Kunar Province – where insurgents have initiated a number of attacks to undermine recent security gains in Afghanistan. While Pakistan's operations are acting as the "hammer" on their side of the border, combined Afghan and ISAF forces are poised to defeat displaced insurgents, acting as the "anvil." Afghan Border Police and other combined security forces are manning outposts along the border and armed drones and close combat aviation are monitoring previously-identified mountain passes that insurgents will likely use as they seek sanctuary in Afghanistan.

Congressional Support

Multi-year security assistance is critical to our efforts in Pakistan. We appreciate continued Congressional support for the Pakistan Counterinsurgency Capabilities Fund, which serves as a key enabler of the Pakistan's military operations against extremists. The fund also provides for a range of partnership activities with potentially transformational long-term effects on our relationship with Pakistan if they can be sustained.

Countering Iran's Destabilizing Activities

Iran's Destabilizing Activities

In view of Iran's destabilizing behavior and its persistent pursuit of a nuclear weapons capability, the Iranian regime's current stance represents the greatest long-term threat to the region. Iran continues to rebuff efforts for engagement, further alienating and isolating itself from much of the rest of the region and from much of the international community. The actions of Iran's leadership squander the potential of its own educated populace and sacrifice the free exchange of ideas for the short-sighted interest of preserving an increasingly harsh and oppressive regime. Recently, Tehran equated the Egyptian protests to the 1979 Islamic Revolution in Iran, making a fanciful and wholly false connection.

The Iranian regime relies on the Islamic Revolutionary Guard Corps-Qods Force (IRGC-QF) to extend influence and create instability across the region through persuasion, coercion, aggression, and targeted messaging. In fact, Iran continues to fund, arm, train, and equip a network of agents, surrogates, and proxies in Iraq, Lebanon, Syria, Gaza, Afghanistan and elsewhere across the region. In the pivotal region of the Levant, Iran seeks to expand its influence, in part by enabling Lebanese Hezbollah and Hamas in order to weaken legitimate governance, limit economic development, and undermine security partnerships. Additionally, Iran delivers weapons and provides military training to surrogates in an effort to target Israel (a nation Iran's leadership have vowed to destroy) and undercut the Middle East Peace Process.

Of urgent concern, the IRGC-QF continues to equip militants in Iraq and Afghanistan that attack U.S. and coalition forces and undermine stability and governance in each of these countries.

The recent January 2011 large caliber improvised rocket assisted mortar (IRAM) attack against U.S. forces in Iraq demonstrated Iran's malicious intent, and ability to escalate violence when they desire.

Iran's Pursuit of Nuclear and Ballistic Missile Weapons

In spite of a fourth round of United Nations sponsored sanctions, Iran appears determined to mature its nuclear weapons program – an ambition that could lead to the proliferation of illicit nuclear materials and spark a nuclear arms race in the region. Admiral Mullen reinforced this point in December 2010, observing:

> "I see Iran continuing on this path to develop nuclear weapons, and I believe that developing and achieving that goal would be very destabilizing to the region." Iran also continues to expand and improve its arsenal of over 2,200 ballistic missiles and long-range rockets, and of approximately 225 fixed and mobile launchers, making it the largest ballistic missile and long-range rocket force in the Middle East. Iran can use these ballistic missiles and rockets, combined with increasing naval capabilities, to threaten global commerce.

Countering Destabilizing Iranian Activities and Keeping Peace with our Partners

Firmly nested within the broader approach of the U. S. government toward Iran, CENTCOM is committed to countering Iran's destabilizing and coercive activities by building confidence with our partners in the region. As one example, we are working together with our Gulf Cooperation Council partners and other nations to advance Integrated Air and Missile Defense. We also conduct activities to reassure our friends in the region that we are with them, preclude conflict, and deter Iran's destabilizing activities, while at the same time standing ready to conduct contingency operations.

Enabling Transition in Iraq

Looking Ahead in Iraq

The year ahead in Iraq presents a significant opportunity for the U.S. to solidify our long-term support to this keystone of regional stability. Our continued investment in Iraq is critical at this juncture, especially given the significant commitment we have made in lives and treasure. Now is not the time to be penny wise and pound foolish with respect to our mission in Iraq. Nested firmly inside the State Department's vision for an enduring U.S.-Iraq strategic partnership, CENTCOM is setting conditions to build on the shared sacrifices between our countries.

The Situation in Iraq

Iraq faces lingering ethnic and sectarian mistrust, tensions between political parties, and strained governmental capacity to provide basic services. Al-Qaeda in Iraq (AQI) remains committed to undermining the Iraqi government and is capable of carrying out orchestrated, high profile attacks. Likewise, Iranian-inspired and equipped proxies continue to be a threat to Iraqi security and governance. While the security situation in Iraq is vastly improved since the peak of sectarian violence there in mid-2007 (violence is currently at all-time lowest levels since 2003), Iraq continues to face significant political, economic, and security challenges. Over the coming year, several factors will determine Iraq's strategic direction, including the continuing development of Iraqi Security Forces (ISF), the effectiveness of the nascent governing coalition, and the degree to which the country is influenced by Iran and threatened by AQI and Shi'a militia elements.

United States Forces-Iraq

From now until the end of this year, United States Forces-Iraq (USF-I) is continuing to partner with ISF during this historic period of transition. USF-I is undertaking a range of activities, foremost among these strengthening the ISF, transitioning security-related activities to Iraq and the U.S. interagency, and contributing to border management and ministerial development.

Establishing OSC-I

Through USF-I and in partnership with the Embassy country team, we are planning the initial stand-up of the Office of Security Cooperation-Iraq (OSC-I) in June of this year and expect it to be fully operational by this October. OSC-I is the cornerstone of our long-term mission to build partner capacity with the ISF. Additionally, the OSC-I will ensure the continuation of the military-to-military relationships that advise, train, and assist Iraqi Security Forces.

Iraq's Regional Integration

Iraq is now at a crossroads, poised to emerge as a positive force for the region after posing security challenges for its neighbors in past decades. Baghdad's selection as the location to host the Arab League Summit is a significant testament to Iraq's re-emergence in the region. Iraq also accepted Egypt's invitation to participate as an observer in CENTCOM's largest exercise, BRIGHT STAR. Jordan has also exerted considerable positive influence in Iraq, training over 1,500 Iraqi Army officers, a number of Iraqi Air Force pilots, and posting a Jordanian defense attaché in Baghdad, in addition to hosting a program to provide extensive training to Iraqi police. Additionally, Jordan, Saudi Arabia, and Kuwait have aided the economic reintegration of Iraq into commercial activity and regularly scheduled transportation networks. Finally, the United Arab Emirates have trained Iraqi police officers in a joint program with Japan and Germany.

Iraq's constructive integration into the region will also help blunt destabilizing Iranian influence. If left vulnerable to Tehran's meddling, Iraq's sovereign future would be imperiled. At the same time that Iran reconstructs shrines, provides electrical power, and constructs schools and clinics in Iraq, Iran also undermines Iraqi political processes, facilitates violence against innocent Iraqi civilians, and provides lethal support to extremist groups targeting U.S. forces. For the U.S. and the international community, a sovereign Iraq under a stable and inclusive government is fundamental to regional stability.

Congressional Support

The support of Congress is critical to facilitating an effective transition in Iraq and in setting the conditions for an enduring U.S.-Iraq partnership. We seek Congressional support in obtaining the appropriate authorities in FY11 to begin immediate facility and site work for the OSC-I to reach full operating capability by October 2011. This is an area of critical need as we work to meet our aggressive timelines. The Iraqi Security Forces Fund critically enables Iraq to set a foundation for its internal and external defense capabilities and provides Iraqi Minister of Interior police forces the training and equipment necessary to maintain internal security without assistance from the Ministry of Defense. Additionally, the ISFF enables Iraqi Army counterinsurgency capabilities and enhances cooperation between the government of Iraq and Kurdish police forces to ensure the consistency of police training and equipment standards throughout Iraq.

Strengthening Central Asian Partnerships

In Central Asia, CENTCOM is committed to strengthening relationships based on those shared interests and goals that we have in common with the Central Asian States of Kazakhstan, Kyrgyzstan, Tajikistan, Turkmenistan, and Uzbekistan. While our nations seek to improve broader economic conditions, CENTCOM is working with our partners to address the migration of extremists in certain areas of Central Asia and to counter the trade of illicit narcotics and human trafficking. Often these activities are interrelated.

Northern Distribution Network

Over the past two years, the development of a robust transportation network has been the most expansive area of cooperation with our Central Asian partners. Our collective agreements with Kyrgyzstan, Kazakhstan, Uzbekistan, and Tajikistan together constitute a logistical system termed the Northern Distribution Network (NDN) used to supply coalition operations in Afghanistan and taking pressure off the Pakistan lines of supply. This diverse network supports the transit of about half of all sustainment cargo to Afghanistan using a variety of sea, air, and land routes. The remaining supplies are flown directly into Afghanistan, trans-shipped from sealift to airlift, or arrive via surface routes through Pakistan. Ultimately, the development and expansion of the NDN and its associated infrastructure will facilitate long-term economic growth in the region, representing a new opportunity for export of Central and South Asia raw materials and exchange of goods in the international marketplace.

Enhancing the Northern Distribution Network

Future NDN efforts are centered on partnering with certain countries to permit two-way flow of all types of wheeled vehicles and associated repair parts, and to increase shipment of cargo already permitted on the NDN (such as building materials). In terms of airlift, Manas Transit Center in Kyrgyzstan is a key Central Asian location that supports aerial refueling and passenger transit missions.

Building Partner Capacity and Pursuing Cooperative Activities

Cooperation Based on Shared Interests

The investment we make in our military-tomilitary engagement to build the capabilities of our partner nation' security forces is a critical component of the whole-of-government efforts in the region. These cost-effective efforts properly place security responsibilities in the hands of other sovereign governments and help to prevent conflicts and instability. With a long-term perspective, CENTCOM carries out partnership activities designed to build strong security capacity and relationships with our friends in the region.

Training

CENTCOM's training and exchanges with our partners are critical to our regional cooperation. CENTCOM has spearheaded the establishment of several Training Centers of Excellence hosted in partner nations, providing world-class mission-specific training for our allies and partners. Existing Centers of Excellence include an Air Warfare Center and an Integrated Air and Missile Defense Center in the United Arab Emirates (UAE); the King Abdullah Special Operations Training Center in Jordan; a NATO Partnership for Peace Combat Engineering and INTERPOL Counter Narcotics Center hosted in Kazakhstan; and an extensive array of associations with the other countries' Professional Military Education programs. Developing Centers include a NAVCENT Maritime Center hosted in Bahrain; a new Explosives Ordinance Disposal school with future Center of Excellence in Saudi Arabia; a proposed Near East South Asia (NESA) branch Center of Excellence in Bahrain; and the Gulf Region Communications, Computer, Command, and Control (C4) Center of Excellence hosted by the Bahraini Minister of Communications.

Exchanges

CENTCOM manages and conducts focused engagement programs with specific partner nations located throughout the AOR in support of the CENTCOM Theater Security Cooperation Plan. The objective is to understand our friend's views and to strengthen relationships and regional organizations to defeat violent extremist networks or situations that threaten the security interests of the region and the U.S. This includes capacity building. Additionally, CENTCOM Headquarters in Tampa, FL is host to over 193 coalition partners from 58 allied nations who make significant contributions to our efforts, and receive invaluable experience interacting with both U.S. forces and our allies.

Equipping

We also provide equipment and security assistance to our regional partners. These activities are among the most important practical steps we can take to demonstrate CENTCOM's enduring commitment to our partners – and to enable interoperable forces in the fight. I ask for continued congressional support of these efforts, including Global Train and Equip, as well as the many security assistance programs managed by the Department of State, including Foreign Military Financing, Foreign Military Sales, and IMET program. As Admiral Mullen noted in his testimony, our security assistance authorities are inflexible, and process are too cumbersome to effectively address today's security challenges in a timely manner. We encourage ongoing efforts to streamline the Foreign Military Financing process in order to cement training and sustainment relations with our critical partners. Accomplishing our mission at CENTCOM requires that we demonstrate our responsiveness to the requests of our partners when we alone should not carry the increasing costs of defending the international order.

Exercises

The final pillar of CENTCOM's partnership activities is our military exercise program. Exercises bolster interoperability between our forces and those of our partners. Each year, our Component Commands conducts more than 50 exercises with our partner nations in the region, including five overseen by CENTCOM Component Commands.

The Long-Term Value of our Exercise Program

The Combatant Commanders Exercise and Engagement program provides critical support to CENTCOM joint training support, exercise and engagement requirements in support of national-level strategic priorities, readiness, and building partnerships within the AOR. Since the beginning of our operations in Afghanistan in 2001, CENTCOM has seen reductions in our exercise program due to ongoing combat operations within the AOR. As combat operations are completed or reduced, restoring sufficient funding levels is critical to support engagement activities with our partners. Without restored funding levels, CENTCOM could lose the advantages gained from a robust exercise engagement program, affecting future access and presence within the AOR and our Theater Security Cooperation Plan. In the interim, we will work imaginatively to make the best use of our exercise budget.

Disrupting Violent Extremist Organizations across the Region

Terrorists in False Religious Garb

The CENTCOM AOR is home to numerous violent extremist organizations (VEOs) comprising a network that, in its own right, represents a considerable threat to the U.S. homeland, U.S. and Western interests, and our allies in the region. The most significant of these is AQ. AQ seeks to impose its morally bankrupt ideology worldwide, and has regional affiliates across the Arabian Peninsula, in Iraq, the Maghreb, and in Somalia (al-Shabaab), with associates including Tehrik-e Taliban Pakistan (TTP), the Afghanistan Taliban, and Lashkar-e-Tayyiba (LeT). The growing cross-organizational cooperation between VEOs replicates mafia syndicates. The organizational success of VEOs is frequently abetted by operating with near impunity in cyberspace.

Attacking VEOs

Along with our interagency and regional partners, CENTCOM continues to develop and implement theater-wide responses in the cyber and physical domains to disrupt and degrade militant networks. Over the past year, interagency efforts have resulted in designating al-Qaeda in the Arabian Peninsula (AQAP) and TTP as foreign terrorist organizations, obtaining a number of Treasury designations, Justice Department arrest warrants, Interpol notices, and placing over 100 individuals and entities on the U.S. Department of Commerce Denial List. Thanks to Congressional funding, the Defense Department Rewards Program has been used by commanders in Iraq and Afghanistan in Fiscal Year 2010 to capture more than 700 high-value individuals, insurgents and terrorists.

Preventing Security Vacuums

In the long-term, CENTCOM is working as a part of an integrated civil-military effort to prevent security vacuums that foment extremism and provide sanctuary to VEOs.

In Yemen, we have forged a tight bond between CENTCOM and our Embassy team in Sana'a to address the heightened threat of AQAP through long-term counterterrorism capacity-building. AQAP cemented its role as a viable and enduring threat to the U.S. Homeland by following-up the failed attempt to bomb Northwest Airlines flight 253 on 25 December 2009 with the "printer cartridge" parcel bomb plot in late October 2010.

Radical cleric Anwar al-Aulaqi publicly spearheads AQAP's campaign against the West, most notably by creating *Inspire* magazine in an effort to encourage Western-based Muslims and enable "lone wolf" style attacks.

In Lebanon, the Lebanese Armed Forces (LAF) has had to navigate a challenging environment in which it does not yet have the monopoly of violence in much of the country. Our assistance has had substantive impact on the ground to include helping the LAF deploy four brigades to the south since 2006 in support of UNHCR 1701 – taking up space where Hezbollah had been. Additionally we have increased the capacity of the LAF Special Operations Forces that won a hard fought battle in 2007 against the alQaeda affiliated Fatah al-Islam movement in the Nahr al-Bared Palestinian refugee camp in Northern Lebanon. The LAF sustained almost 200 killed and 2000 wounded in this operation. We value our close relationship with the Lebanese Armed Forces officer corps based on mutual respect and confidence. We continue to monitor the government formation process in Lebanon and will need to examine the final composition, policies, and behaviors of the next government before making any decisions regarding our relationship, including security assistance, while recognizing that continued engagement with the LAF is an important step in securing its status as an apolitical, non-sectarian, and professional organization.

In Syria, the regime's continuing support for terrorist organizations prevents CENTCOM from developing a military-to-military relationship and limits the scope of U.S. engagement. Consequently, we view the recent return of a U.S. Ambassador to Damascus as a vital piece of our regional security architecture. We stand ready to support Ambassador Ford's diplomatic efforts to produce a more constructive relationship with Syria however we can, and we urge the Senate to confirm his nomination so that he may continue his important work beyond 2011.

Across the region, Theater Security Cooperation activities work against the ability of Iran and extremist elements to destabilize the region. Absent these programs, there is an increasing potential for security vacuums to arise and open the door to greater influence from Iran or violent actors. Our cooperative efforts with regional partners are essential to the long-term effort to address these threats.

Countering the Enemy's Use of the Information Environment

Our enemies are using every available lever of the information environment to promulgate and reinforce their ideology – and, in league with our interagency partners, CENTCOM is committed to countering the efforts of our adversaries. Our enemies operate within cyberspace (and its associated relevant physical infrastructure) to plan, coordinate, recruit, train, equip, execute and garner support for operations against the U.S., its allies and interests. The recruitment of Umar Farouk Abdullmutallab, the unsuccessful Christmas Day Bomber, demonstrates our adversaries' ability to reach across borders, promote their narrative, and defy traditional military constructs to achieve their objectives. Clearly, in the information age, our military must adapt to this new domain of warfare. We ask for the support of Congress to fund our programs that attempt to counter the enemy in the information domain, just as we need funding to disrupt violent extremists in the physical domain.

CENTCOM Activities in the Information Environment

Consistent with the guidance provided by Secretary Gates last December, we conduct Operation Earnest Voice (OEV), which synchronizes and oversees all of our Information Operations activities. OEV seeks to disrupt recruitment and training of suicide bombers; deny safe havens for our adversaries; and counter extremist ideology and propaganda. Full funding of OEV supports all activities associated with degrading the enemy narrative, including web engagement and web-based product distribution capabilities. The effective engagement of our enemies in cyberspace requires the ability for us to conduct a full-spectrum of traditional military activities against them in that domain, including all aspects of Information Operations and Strategic Communication. We coordinate with the Joint Staff, the Interagency, the Intelligence Community, and our coalition partners to examine the adversary's use of cyberspace and identify techniques, tactics and procedures we can use to counter the adversary in the cyber domain.

Combating Weapons of Mass Destruction (WMD)

Risk of WMD

At CENTCOM, we recognize the serious risk and potentially devastating ramifications of a terrorist group, violent extremist organization, or state actor acquiring, proliferating, or using WMD. The nexus between extremist groups, malign state actors, and WMD remains a critical concern throughout the AOR and presents a clear danger to our partners, allies, and the U.S. homeland. CENTCOM remains vigilant in executing the nonproliferation, counter proliferation, and foreign consequence management pillars of America's National Strategy for Combating WMD.

Countering Proliferation and Combating WMD

Countering the proliferation of WMDrelated material is a fundamental aspect of CENTCOM's overall efforts to combat WMD. In concert with our regional partners, CENTCOM is involved with the interagency effort to curtail the ability of adversaries to finance the acquisition of WMD-related items and to deny malign actors the ability to transport suspect dual-use materials across national borders. To this end, CENTCOM plays a key role in containing Iran's evident drive for nuclear weapons in violation of the Nuclear Non Proliferation Treaty by actively enforcing United Nations Security Council Resolutions that sanction the Iranian regime. CENTCOM also supports the interdiction and counter proliferation framework under the Proliferation Security Initiative (PSI). CENTCOM's mainstay program for Combating WMD engagement is the Cooperative Defense Program (CDP). The CDP provides a series of bilateral and multilateral engagement activities to improve U.S. and partner nation interoperability while strengthening partner nations' combating WMD capabilities.

Countering Piracy

The Real and Growing Threat of Piracy

Somali-based pirates continue to prey upon international shipping in the Gulf of Aden, Red Sea, and on the high seas well into the Indian Ocean. Pirates are using previously captured vessels as mother ships to conduct successful attacks as far as 1400 nautical miles from the Somali coast. The number of successful pirate attacks has risen from 42 in 2008, to 51 in 2009, to 68 in 2010. Pirates now hold nearly 700 hostages for ransom. Multi-million dollar per ship ransoms ensure piracy remains a lucrative for pirates and others involved in this criminal enterprise.

A Model for International Cooperation

CENTCOM works with international partners to help patrol the region and to work with interagency partners to gain the prosecution of captured pirates (though we currently lack an international legal framework to detain and prosecute pirates). Piracy is a threat to all, and has promoted international military cooperation that serves as a model for cooperation in other areas. We acknowledge, however, that military action is only one part of the solution, but an essential element nonetheless. NAVCENT coordinates the efforts of over 25 contributing nations to combat piracy at sea and coordinates with European Union Task Force ATALANTA and NATO Standing Naval Maritime Group in Operation OCEAN SHIELD. Pakistan is currently in command of Combined Task Force 151, the international coalition to combat piracy. NAVCENT also hosts a monthly Shared Awareness and De-confliction (SHADE) conference in Bahrain to foster multi-national cooperation and to encourage maritime industry to adopt best practices to defend vessels against piracy. In addition to Coalition, NATO, and EU representation, the conferences also include civilian maritime organizations, and delegates from China, Russia, Japan, and India.

III. Strategic Approach

Many of our challenges are interconnected and require comprehensive long-term solutions, prompting us to adopt an overall approach that is cooperative, integrated, and enduring. As we undertake a diverse range of operations and activities, three principles guide our efforts:

Adopting Cooperative Approaches by Partnering Based on Shared Interests

First, we must adopt cooperative approaches to solving shared challenges. America's strength and security depends on our ability to help our friends in the region defend themselves, underscoring the importance of CENTCOM's initiatives to build partner capacity and pursue bilateral and multilateral initiatives. Starting from our shared interests, we must capitalize on the comparative advantages of all participating nations – for instance, by taking advantage of unique geography or specialized capability. Ideally, such efforts would combine the political, economic, and security spheres of those who choose to participate, strengthening the whole to be greater than the sum of the parts. Our efforts to develop effective solutions for Integrated Air and Missile Defense in the Gulf Region represent a significant example of the kind of cooperative efforts that are necessary to deter and defeat our common threats. As mentioned above, the international coalition to counter piracy in the Somali Basin is a model for multilateral cooperation in the region that not only addresses piracy but also offers opportunities for engagement in other areas.

Our ability to cooperate with our partners depends to a great extent on trust. As a consequence of the confidential diplomatic and military reporting made public by Wikileaks, we must patiently strengthen trust with our partners over time. We are up front with our partners about this episode – which has informed our enemies about supportive leaders as well as our tactics, techniques, and procedures. We remain committed, as ever, to forthright communication in pursuit of our shared objectives. We are reinforcing our efforts to ensure the security of our communications and focusing on enhancing mutually reinforcing objectives with allies and partners.

Integrating Our Efforts by Implementing Civil-Military Solutions

Second, the wars we are fighting today require intensively integrated, comprehensive approaches from the highest to the lowest levels, embracing diplomatic, information, military and economics in an interwoven effort that builds synergy. Promoting security and stability in the CENTCOM AOR cannot be achieved through military means alone. We must therefore look beyond just the traditional application of military power and integrate all elements of national power to address our many challenges. CENTCOM's experience has shown that military might alone is not sufficient to deal with the challenges we confront along with our partners. Diplomacy and Development are just as vital as Defense in securing our national interests. CENTCOM support efforts to address the underlying conditions of instability that fuel current conflicts. Successful application of these instruments of national power, in turn, depends on our ability to achieve harmony within our civil-military relationships. As such, it is a security concern for us when diplomatic posts go unfilled in the region.

The overlapping forces at work in the CENTCOM AOR – those originating from within and outside the region – require exceptional cross-Combatant Command cooperation and coordination. We have achieved progress across AOR geographic seams, exemplified by cooperation with PACOM on matters dealing with China and India and cooperation with EUCOM on Russia, Turkey, and the MEPP. CENTCOM and PACOM regularly synchronize efforts to combat mutual challenges such as piracy, proliferation of WMD, and support to countering violent extremist organizations. Additionally, we continue to work closely with AFRICOM to address the state-failure in Somalia, as well as share critical assets to meet time-critical force requirements. Together we have established a counter-piracy Joint Operating Area in the Somali Basin. We team with U.S. Cyber Command to support global relationships in cyberspace and U.S. Northern Command to protect U.S. borders and domestic security. In all, the cross-Combatant Command effort is going very well.

Supporting Enduring Solutions by Demonstrating Long-term Commitment

Finally, our approach to the region must be enduring. Following through with our long-term commitments in the AOR improves the depth, breadth and quality of our relationships in the region and increases the likelihood of cooperation at the outset. In this region of the world, we are judged by our actions, not words. Individual instances of demonstrated trustworthiness on our part resonate throughout the region for decades. Enduring solutions to the problems that we face also depend on stability, steady economic growth and development in governance. To that end, CENTCOM supports our partners' longterm efforts to grow economically and to develop effective and legitimate institutions of government.

IV. Resourcing the Fight

Beyond the critical funding authorities highlighted above, accomplishing our mission requires that we fully and efficiently resource the following critical enablers. We appreciate Congressional support to provide our warfighters on the battlefront with the tools they need to accomplish their challenging missions. As we adapt to a thinking adversary, we recognize the need to accelerate our acquisition processes to enable us to out-maneuver our enemies. We also recognize the obligation to be good stewards of our nation's monetary resources.

CENTCOM has established stringent control mechanisms to execute our fiscal authorities and to apply the most effective oversight possible of all of our programs.

Intelligence, Surveillance, and Reconnaissance

An Indispensable Tool

There is a considerable and justifiable appetite for ISR capabilities in the CENTCOM AOR. In Afghanistan, persistent ISR capabilities represent one of the most important and effective force multipliers and contribute directly to protecting our troops from the threat of Improvised Explosive Devices through ISR.

In cooperation with the ISR Task Force, we have augmented ISAF forces with a greatly increased capability to counter the Taliban and understand the environment in which we operate. Additionally, as we drawdown our forces from Iraq, we are adjusting the apportionment of ISR in a measured way to ensure that we retain adequate capability to support our force in Iraq while we provide the necessary resources to Afghanistan and elsewhere. We continue to refine our ability to fully integrate U.S. and coalition ISR to deny transnational extremist organizations safe haven, training bases, or staging areas to conduct attacks.

Enhancing ISR Capabilities

We greatly appreciate the support of Congress and the Under Secretary of Defense for Acquisitions and Technology in meeting the ongoing demand for more rapidly delivered ISR collection, exploitation, and dissemination capabilities. Interrelated with our ISR needs, we recognize a need to further enhance integration and synergy between aviation and ground elements that is critical to Combat Air Support and counterinsurgency doctrine. We support a limited objective experiment to refine the requirement for a manned, armed ISR asset attuned to the unique challenges of counterinsurgency in Afghanistan.

Continued investments in ISR technology, infrastructure, architecture, tools, and personnel (particularly trained ISR managers) help us to build on the significant gains we have achieved in the CENTCOM AOR – and enable us to use the arsenal of ISR capabilities currently in the field.

Critical Intelligence Capabilities

Human intelligence and counterintelligence are just as important as technical solutions to remotely gather intelligence, especially in the conduct of operations in wars among the people. Such intelligence activities are inherently government functions that require a long lead time to develop. CENTCOM is posturing for sustained application of our human intelligence capabilities to afford us insights into adversary plans and intentions. CENTCOM is posturing for sustained application of our human intelligence capabilities to afford us insights into adversary plans and intentions.

We are also reshaping our counterintelligence forces to face threats from hostile foreign intelligence services and VEOs that employ sophisticated cyber techniques and trusted insiders to penetrate our networks and compromise our operations.

Improving Force Protection and Countering Improvised Explosive Devices (IEDs)

The Enemy's Weapon of Choice

Now and for the foreseeable future, the enemy is using Improvised Explosive Devices (IEDs) to kill and maim our troops. These devices remain the greatest risk facing U.S. and Coalition forces deployed to Afghanistan and Iraq, as well as a threat to U.S. interests and regional stability throughout the CENTCOM area of responsibility. In Afghanistan, IED attacks account for more than 60 percent of the U.S. and Coalition force casualties, though IED casualties have steadily decreased over the past six months. The flow of lethal aid, migration of IED technology and materials, and development of new tactics techniques and procedures represents a global threat. Homemade explosives, which now account for an estimated 85% of all IEDs, coupled with the proliferation of commercially available IED materials and commercial grade explosives make them relatively cheap and easy to build and employ.

Ongoing Interagency C-IED Efforts

CENTCOM counters the threat of IEDs by working together with all Services and the Joint IED Defeat Organization (JIEDDO). The Services continue to equip U.S. and coalition forces with the latest technology to mitigate and defeat IEDs. Thanks to Congress and the Department of Defense, CENTCOM and our national and international partners have delivered and fielded an unprecedented number of Mine Resistant Ambush Protected family of vehicles throughout Afghanistan. These vehicles have proven critical to safeguarding the tactical mobility of our warriors in harm's way. CENTCOM, in conjunction with the C-IED Senior Integration Group, and JIEDDO have recently fielded a variety of C-IED enablers that have proven to save lives on the battlefield. As a result, we are finding and clearing more IEDs in Iraq and Afghanistan – at a rate above 60 percent for the last 12 months and 70 percent over the last quarter of 2010. These improvements are due in part to more tips from the population, better tactics, and additional enablers, including the effective use of additional ISR provided by the Services to counter this threat.

Attacking the Network

We are going after the entire IED network and insurgent supply lines. Many of our recent successes have come in the use of persistent systems emplaced throughout significant threat areas to help develop insights into the local area. We are concurrently protecting the force using trained dogs, mine rollers, jammers, and handheld devices; the Marines in southern Afghanistan now employ nearly one dog per squad, and soon we will have more than 200 working dogs in Afghanistan. Along with the Services, JIEDDO, and academia we will continue to do everything in our power to ensure our service members and coalition partners have the best technology and training available to defeat the IED threat.

Supporting Additional C-IED Efforts

We continue to call on the defense industry to provide innovative solutions to counter the threat of IEDs. Critical airlift and airdrop sorties dramatically reduce the number of service members exposed to the IED threat. In fact, the number of pounds of supplies airdropped in

Afghanistan has doubled every year since 2005, with an astonishing recovery rate of better than 98 percent. Not all critical movements can be completed by air however. We ask that Congress continue to fund those organizations that provide research and development for the evolution of new and existing counter-IED systems and technologies, especially in areas of pre-detonation, IED stand-off detection, and non-lethal weapons to deny the enemy the ability to deliver or emplace IEDs. We also ask that Congress provide the flexibility to rapidly and proactively counter new, emerging, and future threats that are either present on the battle field or potential threats that represent vulnerability and would be difficult to counter.

Unity of Command and Control of C5 Networks

Coalition, Command, Control, Communications and Computer (C5) networks that meet the challenging demands of our troops in theater are essential to CENTCOM. Currently, the command and control of networks available to our deployed forces is divided among Services, Agencies and Combatant Commands, resulting in degraded and delayed actions that have allowed our adversaries to exploit this fundamental cyber shortfall for too long. One bright network spot, however, is the Afghan Mission Network, which enables U.S. and Coalition forces and civilians to remain connected and synchronized on the battlefield and linked to supporting assets throughout the world. We seek Congressional support to enable effective integration and extension of networks to wherever we fight, from maritime environments to the aerial layer and over rugged mountainous terrain.

Conclusion

In closing, we greatly appreciate the support of Congress on behalf of America's military personnel serving in the CENTCOM region. The stalwart Americans in today's force have been fighting two wars for nearly ten years in the CENTCOM AOR. With remarkable spirit, they look beyond the ambiguity and longevity of today's complex, demanding operations and answer their country's call. Their courage, character and commitment in the face of repeated deployments are inspiring. As their Commander, I am proud to serve alongside them. Thank you very much for your unflagging support of our troops in harm's way and their families here at home.

In: Combatant Commands in the United States Armed Forces ISBN: 978-1-61942-924-6
Editors: Quincy N. Butler and Harris A. Simmons

Chapter 4

TESTIMONY OF ADMIRAL JAMES G. STAVRIDIS, UNITED STATES NAVY COMMANDER, UNITED STATES EUROPEAN COMMAND BEFORE THE 112TH CONGRESS, 2011*

The United States European Command Headquarters

INTRODUCTION

Mr. Chairman, Ranking Member, and distinguished members of the committee, I would like to thank you for the opportunity to appear before you today to share with you the successes achieved and the challenges being faced by the men and women of both the United States European Command and the North Atlantic Treaty Organization's (NATO) Allied Command Operations since I last appeared before you. I have now been at the helm of these commands for almost two years and am happy to report we continue to make progress and develop stronger partnerships for our shared security. The most important activities and initiatives contained in these pages are those in which we work together with our allies and partners to build capacity to ensure U.S. security in the European theater and, thus, defend our homeland forward.

The United States and Europe are inextricably linked—politically as allies and partners in diplomacy. Additionally, the European Union and U.S. economies account for about half the global economy. The two economies are interdependent to a high degree. The United States and the European Union are each other's top trading partners. In 2009, the European Union exported $280 billion in goods to the United States, and imported $220 billion in goods from the United States.

* These remarks were delivered as testimony given on March 29, 2011. Admiral James G. Stavridis, United States Navy Commander, United States European Command before the 112th Congress.

European immigrants entering the U.S. through Ellis Island.

The most important ties for our command, of course, are those between our militaries. U.S. military traditions grew out of European ones. We have learned from each other, often in the demanding circumstances of combat, and we have consistently found ways to become partners and then allies. For the greater part of a century, U.S. soldiers have shared battlefields with their European counterparts – from the Argonne Forest to the sands of Normandy to the mountains of Afghanistan. With respect to mutual and global security perspectives, Europeans are superb partners.

Mission and Vision

The mission of the U.S. European Command is to conduct military operations, international military engagement, and interagency partnering to enhance transatlantic security and defend the United States forward.

We strive to be an agile security organization able to conduct full spectrum activities as part of whole of government solutions to secure enduring stability in Europe and Eurasia.

- Our area of focus covers roughly one-fifth of the planet, including all of Europe, large portions of Asia, parts of the Middle East and the Arctic and Atlantic oceans.
- We are responsible for U.S. military relations with NATO and 51 countries on two continents with a total population of close to one billion people.
- We direct the operation of more than 80,000 military personnel across 10.7 million square miles of land and 13 million square miles of ocean.

- We are responsible for maintaining the quality of life, including health care and schools, for approximately 130,000 Department of Defense family members living in Europe.

Progress

U.S. European Command has achieved progress through proactive initiatives and by responding to challenges and opportunities that arose over the past year.

European Command's Support to International Security Assistance Force (ISAF) in Afghanistan

European Command's activities to support ISAF operations are extensive and effective. At any given time, approximately 80% of the non-U.S. countries deployed to Afghanistan are from the European theater. EUCOM's support to ISAF is largely focused on preparing these partner nations for deployment to Afghanistan.

This includes dispatching mobile planning teams to assess partner nation equipment and training requirements and working with the country to develop a comprehensive pre-deployment plan. These requirements may include provision of equipment such as up-armored high mobility multipurpose wheeled vehicles (HMMWV) and mine-resistant ambush protected vehicles (MRAP), and pre-deployment training to counter improvised explosive devices, build Observer Mentor Liaison Teams, and provide Expeditionary Intelligence Training courses tailored to the complex Afghan counterinsurgency environment.

European Command

Mission

U.S. European Command conducts military operations, international military engagement and interagency partnering to enhance transatlantic security and defend the homeland forward.

Vision

An agile security organization with a "whole of government" approach seeking to support enduring stability and peace in Europe and Eurasia.

Themes

- Ready forces provide regional security.
- Mutual security challenges require cooperative solutions.
- EUCOM is commited to enduring partnerships.

Motto

"Stronger Together"

Combat training for Afghan National Security Forces.

International Security Assistance Force non-commissioned officers (NCO) raise their right hands and repeat the charge of the NCO during an NCO induction ceremony in Afghanistan.

The Command also works closely with our partner nations to fill critical National Training Mission-Afghanistan training requirements.

In quantitative terms, since I last appeared before this committee, European Command has dispatched over 20 mobile planning teams and conducted three Observer Mentor Liaison Team rotations at the Joint Multinational Readiness Center, which included 16 countries and 1,045 personnel and conducted eight Expeditionary Intelligence Training courses in which we have trained 230 personnel from 14 countries. We have also trained over 1,860 soldiers from 15 countries to counter the threats posed by improvised explosive devices, and trained two Polish Brigades and two Georgian battalions for deployment to Afghanistan. On the logistics side, we have moved 487 tons of equipment through Europe to Afghanistan over the Northern Distribution Network.

In FY10, we coordinated use of the Department of Defense Lift and Sustain Program to provide non-reimbursable air and sealift to move 14,897 passengers and 4,206 tons of cargo for 13 contributing nations, who would have otherwise been unable to move equipment and personnel to Afghanistan.

Also, we are able to support partner movements through our active involvement in two major European military transportation consortiums. U.S. European Command also coordinates and schedules the United States' allocation of 1,000 flight hours in the Heavy Airlift Wing (HAW), comprised of three cooperatively shared Hungarian registered and certified C-17 aircraft.

Counter-Improvised Explosive Device Training

A growth area for European Command this fiscal year is counter-improvised explosive device training, where we plan to train as many as 5,000 partner nation soldiers during this fiscal year.

Assuring Access

U.S. European Command plays a critical role in assuring that the United States continues to enjoy access within and beyond European Command's area of focus. Our mature basing footprint includes several locations that are used in support of U.S. Transportation Command's en-route strategy which has proven to be vital in supporting recent operations in Afghanistan and Iraq. Well-established relationships with partner nations further complement our access capability.

Multi-National Joint and Interagency Exercises

The most intensive form of peacetime interaction with our allies and partners occurs in the conduct of joint exercises. European Command maintained a robust bilateral and multilateral exercise program last year, executing 33 major exercises involving nearly 50,000 U.S., allied, and partner nation personnel from 40 nations.

U.S. Army Soldiers from the 173rd Airborne Brigade Combat Team get ready to engage enemy combatants.

The exercises focused on preparing partner nations for ongoing coalition operations including the International Security Assistance Force (ISAF) in Afghanistan, enhancing NATO interoperability, and improving our military capability and interoperability with Israel.

Exercises in the Baltics, Balkans, and Caucasus

In support of NATO, European Command provided forces for nine NATO and NATO Partnership for Peace events in the Baltics. United States Naval Forces Europe also executed Exercise BALTIC OPERATIONS, a long-standing multinational maritime exercise which included 12 nations focused on maritime interdiction and amphibious interoperability. Addressing the Balkans, two major exercises, IMMEDIATE RESPONSE 10 and COMBINED ENDEAVOR 10, bolstered partner capabilities and eased regional tensions. Of particular note, European Command conducted JACKAL STONE 10, a Field Training Exercise in Poland and numerous other locations throughout the world, in cooperation with Special Operations Command. This event involved more than seven nations and approximately 1,100 partner nation Special Operation Forces personnel.

JACKAL STONE, along with other Special Operations exercises and Joint Combined Exchange Training events in over 25 countries, directly supports U.S. and partner Special Operations Forces' readiness and capabilities for U.S., NATO, and European Union missions ranging from counter terrorism to high-intensity conflict.

ARCTIC ZEPHYR

ARCTIC ZEPHYR is a multi-phased exercise, convening initially as a table-top event and projected ultimately to culminate in a multinational search-and-rescue field exercise. In support of U.S. policy and strategy on Arctic issues, the long term goals of ARCTIC ZEPHYR are to maintain an understanding of the legal, commercial, and political ramifications of the changing Arctic environment and to strengthen relationships with other Arctic nations.

This is one of the areas where European Command finds common ground and opportunities for cooperation with Russia.

AUSTERE CHALLENGE

U.S. European Command Headquarters continues to successfully execute the AUSTERE CHALLENGE exercise series, the premier joint force headquarters exercise in the European theater. AUSTERE CHALLENGE 10 forged ahead into new territory when it expanded to train two Joint Task Force Headquarters simultaneously while incorporating a French-led Joint Force Air Component as well as French and Polish Brigade Headquarters response cells, all firsts in European theater training. The benefits of combined, multiple Joint Task Force exercises are clear: challenge the Headquarters and component staffs; reinforce the U.S. position of seeking multi-national solutions; train as we fight; and identify the strengths and limitations of U.S. and coalition interoperability capabilities.

A major advance during AUSTERE CHALLENGE 10 was the establishment of an enduring computer network for future training events and real world operations. European Command planners identified and established the Battlefield Information Collection and Exploitation System as the most capable network for expansion to support Coalition Task Force operations with NATO partners. More importantly, this system is being used at the Joint Multinational Training Center at Grafenwoehr, Germany, by U.S. and coalition forces preparing for deployment to Afghanistan.

AUSTERE CHALLENGE 11 will venture back into the full spectrum, major combat operations arena. Participation will expand outside the theater and there will be a heavy focus on operating in the challenging cyberspace. The 34th Infantry Division Headquarters, from the Minnesota Army National Guard, will provide the Combined/Joint Force Land Component Commander, and French and Polish Land Forces will participate for a second year to operate as Mechanized Brigade Headquarters response cells. We continue building partner capacity by soliciting participation from partner nations around the European theater.

COMBINED ENDEAVOR

During COMBINED ENDEAVOR 10, our premier communications and electronics interoperability exercise, delegates came together from 40 nations (24 NATO and 16 Partnership for Peace countries) to strengthen partnerships, increase communication interoperability, and enhance the capabilities and capacities of partner nations.

This marked COMBINED ENDEAVOR's 16th year building partnerships and featured the participation of Iraq and Afghanistan as observer nations. Both nations committed to increased involvement and a dedication to interoperability between their national forces and NATO/Partnership for Peace nations. COMBINED ENDEAVOR continues to build bridges across Europe and Eurasia and supported the preparation of coalition forces for regional and global operations. In particular, Canada and the United Kingdom tested and trained on the command and control systems they will be using during their upcoming deployments to Afghanistan, and validated operational functionality and interoperability with the same partners with whom they will operate downrange.

Additionally, two new training events were incorporated into COMBINED ENDEAVOR. CYBER ENDEAVOR leveraged the 102d Information Warfare Squadron, from the Rhode Island Air National Guard, to focus on improving the information assurance competencies and network defenses of our European partners, while SHAPE held Exercise STEADFAST COBALT to focus on command, control, communications, and computers in preparation for NATO Response Force deployment and to maximize interoperability with other European nations.

FLEXIBLE LEADER

This year's FLEXIBLE LEADER table-top exercise and senior leader seminar leveraged lessons learned from the Haitian earthquake disaster, and helped validate newly revised plans which European Command planners have written for comparable contingencies. Extensive representation was present from several U.S. Embassy staffs, other U.S. government agencies, and European Command's Service components.

FLEXIBLE LEADER highlighted the extensive work and progress that has been accomplished in foreign consequence management and humanitarian assistance planning, and also identified many courses of action that will improve European Command's ability to respond to a crisis situation quickly and effectively.

American and Italian technicians work to establish network connections during exercise Combined Endeavor 2010.

Patriots to Poland

As stated in the August 2008 Declaration on Strategic Cooperation, and in an effort to strengthen the important strategic partnership between the Republic of Poland and the United States, the U.S. performs quarterly rotations of Patriot Batteries to Poland, enhancing U.S.-Poland air and missile defense cooperation.

These rotations continue to familiarize Polish Armed Forces with the Patriot Missile System and have permitted U.S. forces to share related tactics, techniques, and procedures on missile defense. Of benefit to both U.S. and Polish forces, U.S. Patriot crews have improved their individual tasks and crew drills including operations during deployment, rail activities, and missile transport, storage, and security. Since May 2010, there have been three rotations to Poland for training and exercise purposes only. Although initially focused on one location—Morag, Poland—the last rotation took place in Torun, Poland. A fourth deployment is currently underway.

State Partnership Program

The State Partnership Program accounts for 45% of European Command's military-to-military engagement. Founded in 1993, the State Partnership Program was originally designed to link National Guard states and territories with former Soviet bloc countries for the purpose of fostering mutual interests and establishing long-term relationships across all levels of society.

European Command currently benefits from twenty-one partnerships, we are currently finalizing the protocols for a State Partnership Program with Kosovo, and there is potential for one or two more partnerships over the next couple years.

The true value of this program is the enduring relationships that have been built over time, as many of European Command's state partnerships are approaching their twenty-year anniversaries. Perhaps the greatest example of how critical these relationships are is that, in 2010, National Guard personnel deployed to Afghanistan together with five partner nations as members of Observer Mentor Liaison Teams and other forces training and fighting side-by-side.

European Command Organization

U.S. European Command Headquarters takes a "whole of society" approach to maintaining security and stability while shaping existing structures to adjust to the security environment. These changes will be accompanied by a 15% manpower reduction as we eliminate lower priority missions and identify missions in which European Command can prudently accept additional risk.

At the same time, we are increasing emphasis on emerging mission sets such as ballistic missile defense, military partnering, counter-trafficking, and cyberspace. To further embrace a "whole of society" approach to the security environment, the Command is expanding its J9 directorate to focus on interagency partnering and the use of whole of government/society solutions to strategic challenges.

Hungarian army Sgt. Major Laszlo Pasztercsak instructs U.S. Army Sgt. Mika Putz, of Bismarck, N.D., in the use of the AK63D rifle during a State Partnership Program multinational marksmanship event at Camp Bondsteel.

Additionally, European Command has internally resourced a J7 directorate to provide independent assessments and analyses of strategic and operational processes and products. The Command has also internally resourced a Joint Interagency Counter-Trafficking Center (JICTC)-Europe to focus on the critical counter-trafficking mission across the theater.

We have also established a Military Partnering Center of Excellence to accelerate our efforts to expand the Command's international military partnering engagement activities. The Center will be a virtual, web-based partnering hub for U.S. and European partners to share best practices and lessons learned through military partnering collaboration, networking, and information sharing. The Center will reside on a public web portal linked to existing Department of Defense centers of excellence, most notably the Center for Disaster Management and Humanitarian Assistance, as well as to educational and partnering organizations like the George C. Marshall Center in Garmisch, Germany. We plan to have the Center fully operational by Summer 2011.

Russia Fire Fighting

U.S. European Command also continues to stand ready to provide theater-wide rapid response capabilities for Humanitarian Assistance and Disaster Response missions. This past August, under conditions of severe drought and in the midst of one of the hottest summers on record, a series of destructive wildfires broke out across Russia, prompting President Medvedev to declare a state of emergency in several areas. In response to Russia's request, and in coordination with the U.S. Department of State, EUCOM planners, logisticians, and airmen rapidly identified and airlifted over 36 tons of valuable firefighting supplies and equipment on four U.S. Air Forces in Europe C-130 cargo aircraft to assist Russia in their emergent response to this destructive event. It is precisely this kind of rapid and agile response capability—to a neighboring nation in need—that illustrates the "Smart Power" combination of military capacity and in-stride diplomacy uniquely available to the nation's leadership from its overseas Combatant Commands.

Israel Fire Fighting

Similarly, this past December, a series of wildfires broke out across Israel threatening to engulf lives, homes, critical infrastructure, and valuable natural resources. At Israel's request, and in coordination with the State Department, European Command planners, logisticians, and operators rapidly dispatched five C-130 cargo planes loaded with 60 tons of critical fire-retardant materials necessary to extinguish the blaze. This effort—led by U.S. Air Forces in Europe—played an important role in the international response to Israel in its time of need, and supported our enduring mission to support and promote regional stability and security.

It serves as yet another example of our ability to work together to support each other in times of crisis, demonstrating the value of cooperation among neighbors, allies, and partners, as well as the enduring strength of the U.S.-Israeli relationship.

Efficiencies

While striving to achieve our mission and vision, we are very mindful of today's economic realities. This is why, in support of the Secretary of Defense Efficiencies Initiative, European Command has taken concrete steps to streamline our operations and move toward a more efficient and effective organization. The Command will retain traditional J-staff codes because of their applicability across the Defense Department. However, we are also executing an internal staff rebalance without incurring any growth and leading to a 15% decrease in required manning and budget. We have been taking a close look at all permanent billets and essential mission sets to ensure proper alignment and distribution of resources, and are actively seeking areas where efficiencies may be gained.

> "We must find ways to operate government more efficiently and at a lower cost to taxpayers. Secretary Gates understands the tough economic and fiscal situation facing our nation and I support his efforts in doing everything possible to make every tax-payer dollar count."
>
> Sen. John McCain, Senate Armed Services Committee hearing on DOD efficiencies initiatives, September 2010.

We have focused our mission sets to accommodate our envisioned security environment, based on an analysis of strategic guidance, to include the Unified Command Plan, the Joint Strategic Capabilities Plan, and the Global Employment of the Force. Areas in which we believe we can harvest efficiencies, either through the focusing of missions or by willingness to prudently accept increased mission risk include: personnel management; General/Flag Officer and Senior Executive Service numbers and seniority; General/Flag Officer support; information technology; experimentation; planning, programming, and budgeting system participation; intelligence support; and logistics support. All decisions will be made with careful consideration of their effect on mission readiness.

EUCOM COMPONENTS

Except when conducting joint operations or participating in joint exercises, European Command forces are assigned to, trained, and equipped by our Service-specific headquarters. U.S. Army in Europe (USAREUR), U.S. Marine Forces Europe (MARFOREUR), U.S. Naval Forces Europe (NAVEUR), U.S. Air Forces in Europe (USAFE), and Special Operations Command Europe (SOCEUR) provide the forces for all military-to-military engagements with our partner nations, provide a deterrence function in the region, and serve as deployable units for contingency operations. Understanding these commands is the key to understanding European Command, as they conduct the majority of our day-to-day activities.

U.S. ARMY IN EUROPE
HEIDELBERG, GERMANY

Introduction and Overview

With 42,000 active-duty and reserve service members operating from six enduring Army communities, the U.S. Army in Europe serves as the key coordinator of activities by theater-assigned and rotational ground forces. The integration of these forces across the full spectrum of operations places responsibilities on the Army in Europe ranging from the tactical and operational to the strategic level. With V Corps' intermediate tactical headquarters capability deployed in support of International Security Assistance Force (ISAF) efforts in Afghanistan, the Army in Europe directly oversees capacity building activities throughout European Command's area of focus, including efforts in support of U.S. Africa Command. It also maintains critical logistical lead component duties in support of NATO's Kosovo operations. The Army in Europe translates strategic directives from the Command and the Department of the Army into executable tactical orders for subordinate units, including planning for and executing key tasks in support of our European partners and Israel.

Under these circumstances, the Army in Europe provides key tactical and operational forces, to include full spectrum combat units and strategic enablers, for global employment. As the U.S. military land component provider on the continent, the Army in Europe leads the ground effort in building partner capacity in support of global requirements. These efforts continue to be instrumental in supporting the ISAF Commander's requirement for improving the effectiveness of coalition ground force deployments, by training and preparing U.S. and European forces for operations in Iraq and Afghanistan.

Major Accomplishments

With over 25% of its assigned forces deployed in support of operations in Iraq and Afghanistan, the Army in Europe continued to support the Secretary of Defense's commitment to improving the capacity of coalition partners and allies by conducting a robust training and exercise program designed to build partner capacity and increase coalition interoperability.

Building Partner Capacity

In 2010, the Army in Europe provided 755 soldiers from ten nations with life-saving drivers' training on Mine Resistant Ambush Protected (MRAP) vehicles. It trained an additional 422 soldiers from nine nations to counter improvised explosive devices. Supporting U.S. national security objectives, this training enabled the Command to prepare full- spectrum capable forces for global employment while improving partners' and allies' survivability and effectiveness during deployments. Additionally, in support of the ISAF Commander's top priority to train Afghan soldiers and police, the Army in Europe trained 50 Operational Mentor-Liaison Teams (OMLT) and Police Mentor-Liaison Teams (POMLT) for deployment to Afghanistan.

Critically, the U.S. Army in Europe has worked to advance defense institutional transformation.

One example of this was its assistance to Romania in the development of the country's tactical, operational, and strategic military doctrine. The Army in Europe also promoted U.S./partner interoperability through its joint exercises and exchange programs, recently having conducted leader exchange programs with Israel in the areas of aviation, ground maneuver, training, reconnaissance, and military intelligence. Additionally, the Army in Europe enhanced capacity for maintaining regional stability—evidenced by its support to Poland— executing the first three Patriots-to-Poland rotations.

Croatian and Minnesota National Guard soldiers prepare to clear a room during an Operational Mentoring and Liaison Team training exercise at the Joint Multinational Readiness Center.

Both on and off the continent, the Army in Europe planned for and exercised ground capabilities to support key NATO partners and Israel in defending against potential threats. It has worked with Israel to significantly improve their ballistic missile early warning capability and has enhanced bilateral air and missile defense training exercises in coordination with our partner.

In addition to this assurance to allies and deterrence of potential aggressors, its ability to provide foreign consequence management and foreign humanitarian assistance ensured that the Army in Europe was, and continues to remain, prepared to support the United States' Article IV and V commitments to our NATO partners. Finally, as part of the larger U.S. efforts to reset our relations with Russia, U.S. Army in Europe included Russian military leaders as observers for three major exercises.

Exercises

Exercises continue to enhance the pre-deployment training of U.S. and coalition forces for current contingency operations, and serve to prepare these same forces for future coalition operations. This past year, the U.S. Army in Europe participated in 21 major exercises (including three mission rehearsal exercises) conducted in 11 countries with 28 participating nations. Among these exercises was JUNIPER FALCON 11, an Israeli-led Joint Task Force-level exercise focused on improving Israeli Defense Force command and control and logistical capacity. Supporting America's continuing partnership with Turkey, COOPERATIVE RESOLVE 10 was a battalion-level Command Post Exercise conducted in

Turkey meant to enhance interoperability. As a final highlight, RAPID TRIDENT 10 was a peacekeeping exercise involving 16 countries conducted in Ukraine to support its NATO interoperability goals through NATO's Annual National Program.

Humanitarian Assistance

As part of our Humanitarian Assistance program, the Command also provided enabling support to U.S. Africa Command for MEDFLAG 10, a medical skills exchange exercise conducted in the Democratic Republic of Congo by U.S. and African militaries.

Way Ahead

The U.S. Army in Europe will provide combat power to support global operations while at the same time continuing to build partner capacity. Together, these efforts will help advance the long-term process of defense institutional transformation of U.S. partners and allies.

As the Army in Europe trains and prepares for deployment alongside coalition partners in the coming year, it will continue to foster the residual effects of increased training effectiveness among friendly militaries—from improvement in counterinsurgency operations and incorporation of interagency concerns, to the updating of U.S./NATO interoperability doctrine. Specifically, in support of the President's introduction of the European Phased Adaptive Approach, the U.S. Army in Europe will train with and provide assistance to European forces to ensure their ability to defend against ballistic missile threats. As part of a separate mission, the U.S. Army in Europe will provide similar assistance to Israel through bilateral agreements. As a whole, the engagements by the Army in Europe will continue to focus on improving relations, enhancing interoperability, focusing on mission success of both in- and out-of-area operations, and ensuring access to facilities and infrastructure throughout the theater.

> "Where possible, our strategy is to employ indirect approaches – primarily through building the capacity of partner governments and their security forces."
>
> Secretary of Defense Robert Gates.

Marine Forces Europe
Stuttgart, Germany

Introduction and Overview

In 2010, Marine Forces Europe, with approximately 150 personnel assigned, focused on building partner capacity through combined activities and utilizing expeditionary forces to reassure allies, deter potential adversaries, and remain ready to respond rapidly to crises in the region. The U.S. Marine Corps' expeditionary nature drives Marine Forces Europe to a primary orientation on security cooperation activities with our newest NATO allies and partners in the Caucasus, Black Sea, Balkan, and Baltic regions.

Major Accomplishments. With only a small service component headquarters, Marine Forces Europe very effectively leveraged the capabilities of the Marine Corps in support of European Command objectives. Marine Forces Europe's activities focused on building partner capacity to contribute to collective security in Europe and out-of-area operations, such as support to the International Security Assistance Force, through targeted security cooperation and combined exercises.

Building Partner Capacity

Georgia Deployment Program-International Security Assistance Force: This U.S. Marine Corps-led joint program is successfully training and deploying Georgian infantry battalions to fight alongside NATO forces in the volatile Helmand Province in Afghanistan. Through an intense partnering concept with the Marines, the Georgian Armed Forces have significantly increased their institutional capacity to plan and conduct training for units preparing to operate in a full spectrum counter-insurgency environment.

USMC Black Sea Rotational Force: During the summer of 2009, the U.S. Marine Corps provided a Special Purpose Marine Air—Ground Task Force to conduct security cooperation in support of European Command's theater objectives. Forward deployed and operating out of the temporary Task Force East facilities at MK Airfield in Romania and the Novo Selo Training Area in Bulgaria, the Black Sea Rotational Force conducted numerous and diverse targeted multi-national security cooperation activities with 12 partner and allied nations in the Black Sea, Balkans, and Caucasus regions to enhance partner military capabilities, expand U.S. and NATO access to strategic regions, and promote regional stability.

Exercises

In 2010, Marine Forces Europe, in coordination with U.S. Naval Forces Europe, effectively reassured allies and deterred potential adversaries by exercising combined maritime expeditionary capabilities and improving European Command's and NATO's ability to rapidly deploy and assemble expeditionary forces in the region during several historic exercises. In total, Marine Forces Europe participated in 13 exercises to include joint, multilateral, and bilateral exercises in 2010.

COLD RESPONSE 10 was a Norwegian-hosted 14-nation exercise conducted north of the Arctic Circle and focused on maritime/amphibious operations and interoperability. U.S. Marines, under the Tactical Control of the United Kingdom's 45 Commando Battalion of the Royal Marines and embarked on Her Netherlands Majesty's Ship Johan De Witt, participated in a brigade-sized beach assault.

As part of BALTIC OPERATIONS 10, U.S. Marine and U.S. Navy forces, along with our Baltic State allies, conducted both a Maritime-Prepositioning Force offload and onward movement of combat equipment in Ventspils, Latvia, and a combined amphibious landing in Estonia during this European Command-sponsored Partnership for Peace Exercise. The overwhelmingly positive response by our Baltic allies once again validated the unique and critical role of Maritime-Prepositioning Forces' ability to rapidly respond to crises and support our allies.

Humanitarian Assistance

As part of the Marine Corps Prepositioning Program— Norway, Marine Forces Europe works with the Norwegian Defense Staff, European Command, and Headquarters, U.S. Marine Corps, to develop and refine plans that enhance access to prepositioned equipment ashore for U.S./NATO operations and crisis response, Theater Security Cooperation, and Humanitarian Assistance/Disaster Relief.

Way Ahead

Although it achieved successes with its current force posture, Marine Forces Europe will continue to seek greater Amphibious Ready Group/Marine Expeditionary Unit presence to satisfy the consistent demand by global core partners such as the United Kingdom and France for bilateral combined-arms and amphibious training. This type of training has largely been absent in the European Command theater since 2003. Resuming a sustained presence in the European Command region would serve to deter adversaries and assure allies and partners of our commitment to stability in Europe.

Marine Forces Europe will continue to build partner capacity in theater through ongoing support to the Georgia Deployment Program-International Security Assistance Force and the Black Sea Rotational Force, deploying in 2011 to support scheduled engagement and security cooperation activities with 14 partner nations. The Black Sea Rotational Force is expanding its activities to include preparing partner nations for deployment to Afghanistan and conducting Non-Commissioned Officer development.

U.S. Marines amphibious assault vehicles embarked aboard the USS Gunston Hall (LSD 44) depart the well deck during the start of a simulated amphibious assault. The combined U.S., Estonian amphibious landing was part of a demonstration of the interoperability between the two forces.

U.S. Naval Forces Europe
Naples, Italy

Introduction and Overview

With approximately 8,000 active-duty and reserve services members operating from four main installations supporting rotational air, surface, submarine and expeditionary forces, U.S. Naval Forces Europe conducts the full range of maritime operations and Theater Security Cooperation in concert with NATO, coalition, joint, interagency and other partners in Europe. Naval Forces Europe continues to perform Navy Component Commander functions which support day-to-day fleet operations and Joint Maritime Commander/Joint Task Force Commander missions in support of European Command. Its presence not only strengthens relationships with enduring allies, it also develops maritime capabilities with emerging partners, particularly in the European Command's southern and eastern regions. Naval Forces Europe is enhancing maritime security in these regions through the development of maritime domain awareness, trained professionals, maritime infrastructure, response capabilities, regional integration, and a comprehensive approach for planning and execution.

Major Accomplishments

In 2010, Naval Forces Europe met all warfighter mission requirements and maintained certification as Joint Force Maritime Component Commander, Europe. Additionally, the component focused energy and resources on Theater Security Cooperation activities to enhance interoperability between allies and the maritime capabilities of partner nations. These activities developed partner capacity through multiple events, exercises, and operations in order to promote maritime domain awareness, security, and sea control. Always ready to respond to crises in the European theater, Naval Forces Europe also contributed to global efforts, such as Operation Enduring Freedom and anti-piracy operations off the coast of Africa by delivering trained forces, strengthening international relations, and increasing the efficiency of our interactions with our allies and partners.

Theater Submarine Operations

The stable presence and patrolling of U.S. Submarine Forces in the European theater defends U.S. national security forward, even as it enhances the security of our allies and key partners. Our submariners are engaging in vital missions that contribute directly to European Command's core missions of transatlantic security and building partnership capacity. U.S. submariners have fully leveraged deployment time in theater, honing their skills and contributing to the Command's capacity to conduct critical intelligence, surveillance, and reconnaissance operations, anti-submarine and anti-surface warfare, undersea dominance, close proximity strike, high-value unit protection, and free and unfettered access to the vital sea lines of communication. In building our partners' capacity, European Command has also facilitated the participation of Allied submarines, submariners, and associated staff in a vigorous series of theater Anti-Submarine Warfare exercises with U.S. submarines, exercises that all parties have leveraged to improve their warfighting proficiency and – importantly – their interoperability.

These factors and opportunities are increasingly important as we observe a highly capable Russian submarine fleet whose pace, scope, and sophistication have risen dramatically in recent years. The Russian Navy has four new classes of submarines in development or near delivery. The next-generation KILO-class submarine—the ST. PETERSBURG class—is nearing completion. Available for export, it represents a significant improvement in both capability and quieting. The European Command area of focus is also the stage for the most sensitive Russian submarine operations and advanced weapons testing. Just this past October, Russian submarines successfully fired three submarine-launched ballistic missiles over a period of two days. These operations reveal a renewed Russian focus on the undersea arena.

Members of the visit, board, search, and seizure team from the guided-missile destroyer USS Forrest Sherman (DDG 98) return to Forrest Sherman after participating in a boarding training exercise during BALTIC OPERATIONS.

Intelligence, Surveillance and Reconnaissance

Naval Forces Europe expanded European Command's intelligence, surveillance and reconnaissance in support of allies through Operation NOMAD SHADOW, Operation SPRING OFFENSIVE, and the first-ever ship-based Unmanned Aerial Vehicle missions in the Black Sea. It also expanded the capability of naval bases in Rota, Spain, and Sigonella, Italy, to support intelligence, surveillance, and reconnaissance assets in support of both the European and Africa Commands.

Building Partner Capacity

Partnership and security cooperation events conducted by Naval Forces Europe include:

Eurasia Partnership Capstone

Naval Forces Europe's flagship initiative throughout the Black and Caspian Sea regions was designed to integrate disparate regional efforts by all maritime partners into a global maritime partnership for Eurasia. This year's October event drew 110 senior enlisted and junior officer attendees from nine partner nations: Azerbaijan; Bulgaria; Georgia; Greece; Lithuania; Malta; Poland; Romania; and Ukraine.

Courses were held at the Maltese Navy Training Facilities. Topics included: maritime interdiction operations; visit, board, search, and seizure procedures; search and rescue procedures; maritime law; and environmental protection.

USNS GRAPPLE

In August, a Navy auxiliary salvage ship and embedded Mobile Diving and Salvage Company removed and scuttled six sunken and decaying patrol boats in Sarandë, Albania. These boats were towed to another location near the harbor and used to make an artificial reef, which is expected to boost the local economy through tourism. More importantly, the pier at the auxiliary naval base is now clear and can be handed over for use by the local fishing fleet, reducing unnecessary defense overhead involved with support of this area.

Exercises

In 2010, Naval Forces Europe participated in nine Joint Chiefs of Staff exercises and numerous NATO and European Command exercises. BALTIC OPERATIONS 10 involved 11 European and NATO nations (including Russia), 32 ships, 200 vehicles, and over 3,000 personnel across a 1,000-square kilometer operating area in the Baltic Sea region. This annual exercise promotes mutual understanding, cooperation, confidence, and interoperability among forces and personnel of participating nations. SEA BREEZE 10 was co-hosted by Ukraine and the United States, involving11 European nations, 24 ships, 13 aircraft, and over 2000 personnel across the Black Sea and Ukraine. This exercise enhanced the maritime capability of Black Sea and Partnership for Peace nations by exercising collective maritime safety, security, and stability actions. BREEZE 10 was co-hosted by the Bulgarian Navy, involved six European nations, and culminated in the NATO Response Force certification of Bulgarian and Romanian ships.

Humanitarian Assistance

Naval Forces Europe provided platforms, personnel, and resources to support Humanitarian Assistance/Disaster Relief for Russian wildfires, the Ethiopian Airlines crash off the coast of Lebanon, and flooding in Portugal and Albania. The U.S. Navy routinely provides foreign humanitarian assistance in times of crisis, in cooperation and collaboration with many nations around the world.

Way Ahead

While responding to crises in the European theater will take precedence over all other activities in 2011, Naval Forces Europe will also continue to build the maritime capacity of our allies, as well as new and emerging partners in a cost-effective and responsible manner with the desire of advancing all parties toward full interoperability and participation with our forces. Naval Forces Europe will continue to improve information assurance and cyber system security of Command and Control/Information Systems. Interoperability with NATO Allies and achievement of NATO standards by new NATO partners, including preparing STRIKFORNATO for the NATO Response Force 2012 rotation, will be important goals. The ballistic missile defense capability, both afloat and ashore, will continue to develop and expand. Our forces will focus on the priorities of maritime safety, security, cooperation, and crisis response with the overall goal of advancing U.S. interests in the region.

U.S. Air Forces in Europe
Ramstein Air Base, Germany

Introduction and Overvie

With just under 26,000 active-duty, guard, and reserve service members operating from five main operating bases supporting nine wings and many geographically separated locations, U.S. Air Forces in Europe is a key force provider of forward-based, full-spectrum airpower in support of European Command and Defense Department objectives. This posture enables simultaneous support to ongoing global operations, ensures global strategic access, assures allies, deters aggression, and remains the key to building partnerships.

Major Accomplishments

During 2010, Air Forces in Europe supported ongoing contingency operations and worked daily with our NATO allies and partners to ensure security in the European theater and defend our homeland forward. Supporting contingency operations, 2,800 Air Forces in Europe personnel were deployed at any given time throughout 2010. In Afghanistan, our fighters flew 23,500 combat hours, representing 39% of the total U.S. Air Force fighter hours flown. In addition, one of two Control and Reporting Centers assigned to Air Forces in Europe was continuously deployed, providing constant air battle management and control. We contributed vital data links for world-wide communications, unmanned aerial system command and control, intelligence collection, and space operations. U.S. Air Forces in Europe supported the treatment and movement of over 12,000 patients to and from Landstuhl Regional Medical Center. Finally, Air Forces in Europe continued to maintain critical en route infrastructure.

This system of bases supported global air operations to three geographic combatant command theaters and smooth transit of over two-thirds of air mobility missions.

Supporting contingency operations at the same rate as U.S.-based forces, Air Forces in Europe simultaneously conducted European Command operational requirements. In addition to flying Combat Air Patrols in support of NATO's Baltic Air Policing mission, Air Forces in Europe continue to conduct intelligence, surveillance, and reconnaissance missions in the greater Levant region, as well as participate in the Georgia Deployment Program.

A Polish Air Force F-16 receives fuel from a U.S. Air Force KC-135 over the skies of Latvia during a NATO Baltic Air Sovereignty Training event.

In support of the Phased Adaptive Approach for Missile Defense in Europe, U.S. Air Forces in Europe led a U.S. and NATO Ballistic Missile Defense Task Force. This Task Force developed a concept of operations which significantly contributed to NATO's recent decision to adopt this critical mission. U.S. Air Forces in Europe began laying the foundation for ballistic missile defense integrated command and control architecture in line with NATO Summit agreements, and led U.S. and NATO data-sharing integration efforts, which resulted in the development of clear strategic ballistic missile defense requirements and a demonstrated capability to exchange information between U.S. and NATO systems.

Building Partner Capacity

When not supporting combat operations, U.S. Air Forces in Europe units serve in a permanent role of building partnerships and partner capacity. In 2010, Air Forces in Europe conducted 767 building partnership engagements with 39 participant nations. Two-thirds of these events contributed to partner interoperability for Afghanistan operations.

Nearly 100 events in the Baltics, Poland, Bulgaria, and Romania focused on interoperability of NATO standards and equipment. As a result, these particular events fostered effective mobility operations and ensured strategic access to ranges, airspace, and airfields. In Poland, Air Forces in Europe units helped develop capability to deploy and employ Polish F-16 and C-130 aircraft. Air Forces in Europe conducted numerous tactical exchanges with Polish F-16 and C-130 pilots, providing training in all manner of combat and air mobility operations. As a result of these efforts, the Polish Air Force is advancing towards its goal of passing its first NATO Tactical Evaluation in 2011. In Romania, U.S. Air Forces in Europe units assisted Romanian units across the full spectrum of air mobility operations. Both nations jointly exercised tactics and procedures to improve airfield planning and operations, combat search and rescue, and aircrew management procedures to improve Romanian air capabilities. As a result, the Romanian military is now a self- deployable force using its C-130s to move to, and conduct operations in, Afghanistan. Our engagement efforts also highlight the intrinsic value of non-commissioned officer development. Each of our in-country engagements highlights the need for empowered junior officers and non-commissioned officers, who serve as the backbone of effective military operations.These efforts serve to improve the defensive development of our allies and partners, and directly improve interoperability of forces engaged in combat. U.S. Air Forces in Europe trained with 28 nations to provide over 100 Joint Terminal Attack Controllers, highly-trained Airmen providing the key link between airborne assets and supported ground combat units. As a result, Joint Terminal Attack Controllers from 15 nations now fight alongside their American counterparts in Afghanistan. Furthermore, training at U.S. Air Forces in Europe's Warrior Preparation Center provides tangible and continued assurance of the U.S. commitment to allied security while developing capabilities actively employed in International Security Assistance Force operations.

Exercises

In 2010, U.S. Air Forces in Europe participated in 60 Joint Chiefs of Staff, NATO, and European Command exercises. Ten nations, including Armenia, Azerbaijan, Georgia,

Macedonia, Serbia and Ukraine, exercised to develop peace support operations, provide medical training, and foster interoperability. Likewise, 40 nations from North America, Europe, and the Middle East exercised to improve computer network and communication capabilities necessary for effective responses to natural disasters. During Exercise BRILLIANT ARDENT, 60 aircraft from six nations, operating from bases in Germany, the Czech Republic, France, Poland, and the United Kingdom, validated the responsiveness and capabilities of the NATO Response Force.

Humanitarian Assistance

U.S. Air Forces in Europe answered the call for assistance several times in 2010 with humanitarian airlift. This summer, we delivered over 36 tons of equipment to fight wildfires in Russia. In December, we airlifted 60 tons of fire retardant materials for wildfire relief near Haifa, Israel. In one of our building partnership capacity success stories, the Heavy Airlift Wing at Papa Air Base, Hungary, executed several humanitarian relief missions. During Operation UNIFIED RESPONSE, Heavy Airlift Wing C-17s delivered 34 Swedish aid workers and 135 tons of aid to Haiti. In July, following the devastating flooding in Pakistan, the Heavy Airlift Wing delivered over 42 tons of medical supplies to Karachi, Pakistan. In April, the Wing repatriated the remains of the victims from the Polish air tragedy near Smolensk, Russia. Through continued training and commitment, the 12 nations of the C-17 Heavy Airlift Wing flew over 2,800 mishap-free hours in its second year of existence.

Way Ahea

As we move forward into 2011, U.S. Air Forces in Europe will continue its focus on operating efficiently and effectively, maximizing its strategic location to support current operations while simultaneously expanding its building partnership portfolio. The Command will continue to support the President's Phased Adaptive Approach for Missile Defense in Europe. Working closely with NATO, Air Forces in Europe will deliver an Operational Level Concept, develop requirements for effective information sharing with allies, and establish the necessary venues for effective education and training in this mission area critical for the security of the United States, our allies, and partners.

U.S. Special Operations Command Europe
Stuttgart, Germany

Introduction and Overview

Special Operations Command Europe is a joint command comprised of more than 1,500 active-duty and reserve personnel operating from two main forward-deployed locations: Stuttgart, Germany and Mildenhall, England. It has three assigned components: 1st Battalion, 10th Special Forces Group (Airborne); Naval Special Warfare Unit-2; and the 352d Special

Operations Group. Special Operations Command Europe continues to contribute significantly to the development of partner special operations forces and stands ready to defend against transnational threats and rapidly respond to unforeseen contingencies within the European theater.

Major Accomplishments

In 2010, Special Operations Command Europe remained heavily engaged, conducting 25 joint combined exchange training events, six bilateral training activities, 46 Partnership Development Program events, and two bilateral counter-narcoterrorism training events. The Command augmented this effort with numerous key leader engagements and staff visits to further develop partner special operations forces' organizational, institutional, and staff capabilities. Finally, Special Operations Command Europe deployed special operations forces company-sized elements to both Afghanistan and Iraq, as well as smaller Joint Planning and Advisory Teams and staff augmentation to support partner special operations forces in Afghanistan. Focus areas for 2010 included building partnerships, supporting operations in Afghanistan, and countering transnational threats, all of which directly support European Command's core mission of international military and interagency partnering to enhance transatlantic security and defend the homeland forward.

Building Partner Capacity

The Command's effort to build partnerships focused on enhancing allied and partner special operations forces' interoperability and validating capabilities through the Partner Development Program and multilateral special operations exercises. The Partner Development Program is focused on training partner and allied special operations forces in military assistance operations and is a proven strategy for building the capacity of allied and partner special operations forces capacity, primarily for the International Security Assistance Force. With adequate sustainment, it will also provide long-term special operations forces' capacity for future conflicts and out-of-area operations. Since its inception in 2007, and its initial funding as a program of record in 2009, European special operations forces' participation in the International Security Assistance Force has increased nearly 500%. Poland, Romania, Hungary, Lithuania, and the Czech Republic have increased their investment and commitment of special operations forces in Afghanistan with exceptional results. With military assistance being the primary special operations mission in Afghanistan today, and the continued focus on developing Afghan security forces, participation in this program will likely remain an important mission in the future.

Over the past year, the Partner Development Program has also focused on increasing collective special operations force rotary-wing aviation capacity among our allies and partners. Special operations force rotary-wing assets, deployed in support of the International Security Assistance Force, will continue to be a high-demand requirement for future contingency operations. Training allied and partner aircrews and helping to upgrade their airframes, such as the MI-17, is an efficient solution for increasing rotary-wing capacity and capability to support contingency operations. For example, the cost to upgrade five MI-17s and train the crews and maintenance personnel to the basic standard required to support special operations forces is roughly equal to the cost of one new U.S. CH-47G helicopter, not including training or maintenance. Additional helicopter capacity is not only efficient; it also decreases the deaths and serious injuries caused by improvised explosive devices, by limiting road movements and increasing medical evacuation capacity. Thus, additional helicopter capacity could lead to potential increased troop contributions among these nations.

While successful, the Partner Development Program has not reached its full potential. The program is hampered by cumbersome resourcing processes. Lack of focused special operations forces resourcing to create a strategic capability makes it difficult to build enduring special operations forces capacity.

Exercises

Special Operations Command Europe supported the annual international theater-wide special operations forces capstone exercise, JACKAL STONE 10. This year's exercise was hosted by Poland and Lithuania and brought together approximately 1,100 special operations forces service members from seven nations: Poland, Lithuania, Ukraine, Latvia, Croatia, Romania, and the United States. As a key element of the Partnership Development Program, JACKAL STONE provides special operations forces the opportunity to train together and build mutual respect while sharing doctrinal concepts, training concepts, and various tactics, techniques, and procedures.

Coalition special forces wait for the MH-47G Chinook to land so they can extract their high value target during the opening ceremony for Jackal Stone 10.

Support Operations in Afghanistan

Since 2007, Special Operations Command Europe has supported the International Security Assistance Force through the continued deployment of a U.S. Army Special Forces Company and three U.S. Navy SEAL Joint Planning and Advisory Teams, which deploy and operate with our partner special operations forces. We also continue to provide staff officers as rotational augmentees to the NATO Special Operations Forces Fusion Cell, a strategic element for building interoperability among partner special operations forces.

Way Ahead

Looking to 2011 and beyond, Special Operations Command Europe will continue to increase allied and partner special operations forces' capabilities and capacity, both on the ground and in the air. It will seek to increase strategic understanding and partnerships as the European Distributed Special Operations Forces Network develops. Special Operations Command Europe will also continue its close partnership with the NATO Special Operations Headquarters and its International Security Assistance Force partners. Support and enablement of International Security Assistance Force's special operations forces will continue to be a high priority, as will efforts to counter transnational threats in order to protect our homeland.

CHALLENGES AND OPPORTUNITIES

In a theater as large and dynamic as U.S. European Command's, the opportunities to initiate positive change are abundant, as are the ways in which security can regress unless we remain alert and attentive to negative trends. Every challenge in the region presents an opportunity for engagement and cooperation.

Afghanistan

Of the 49 nations besides the United States that have contributed 45,000 forces to the International Security Assistance Force, approximately 80% of them (37 nations) come from the European theater. Together, these 37 nations have contributed nearly a third of the military personnel serving in Afghanistan. And they have suffered, with hundreds killed in action. Supporting the International Security Assistance Force has given European Command the opportunity to deepen its relationships with our allies and partners, using our expertise and experience to inculcate an expeditionary mindset and train deploying partner nation forces in irregular warfare.

The contributions and sacrifice of Eurasian and European nations in Afghanistan have demonstrated the credibility, legitimacy, and effectiveness of our international military cooperation. The scale of allied and partner force contributions to the International Security Assistance Force has allowed the hand-over of significant responsibility for regional operations to coalition partners. NATO's Operational Mentoring and Liaison Team program directly supports the development of the Afghan National Army, and the Police Operational Mentoring and Liaison Teams program supports the expansion of the Afghan National Police. Under these programs, European allies and partners are currently providing approximately 50% of the number of teams required to train Afghanistan's security forces.

> "Having Afghans lead these operations is the Taliban's worst nightmare, because it gives the lie to the Taliban propaganda that portrays Western troops as hostile occupiers. Afghans themselves will be more effective than our troops in winning the trust of the Afghan people."
>
> Sen. Carl Levin, Chairman of the Senate Armed Services Committee, at the Council on Foreign Relations, October 2010.

Additionally, at any point in time, there are approximately 10,000 U.S. personnel, assigned to European Command, that are deployed to Afghanistan making vital contributions on a daily basis. However, within the European theater itself, European Command's focus is to support other nations as they seek to contribute to security and stability efforts in Afghanistan.

Many nations are making particularly large force contributions and have suffered high casualty rates relative to their populations. Our partners understand the importance of this mission, and they are willing to send their sons and daughters in harm's way alongside our own to bring peace, security, and prosperity to the people of Afghanistan. Within the European theater itself, European Command lends whatever support it can to these nations as they seek to contribute to security and stability efforts in Afghanistan. Within the framework of contributing to international efforts in Afghanistan, and within the boundaries and authorities set by law, regulation, and international agreements, this support involves providing training, equipment, logistical assistance, and personnel augmentation.

The Balkans

The United States' continuing support to NATO's Kosovo Force and Operation JOINT GUARDIAN helps maintain stability in Kosovo and advances security progress alongside our NATO and European Union partners. European Command has participated in NATO operations in Kosovo since 1999. European Command supports Kosovo Force through our land component, U.S. Army in Europe, and leverages National Guard Bureau forces to source mentors and advisors for Task Force Falcon (Multinational Task Force-East), Regional Mentoring and Liaison Teams, NATO Training Teams, and elements of the Kosovo Force Headquarters, as well as to augment the Kosovo Force Military-Civilian Advisory Division. NATO presence was reduced from a peak of 14,000 in 1999 to 10,000 in January 2010, when it began reductions to a strength of 5,000. Kosovo remains stable and secure, as demonstrated in the peaceful conduct of elections in December 2010.

European Command has played a significant role in Bosnia's progress since the 1995 implementation of the Dayton Accords. At the height of Operation JOINT ENDEAVOR in 1996, more than 20,000 U.S. service members served in Bosnia. The September 2009 deactivation of Task Force Dayton, the last U.S. entity operating in Bosnia, marked a significant milestone for European Command. Less than forty U.S. personnel now remain in Bosnia assigned to the NATO Headquarters-Sarajevo and the United States Balkans National Support Element. European Command continues building partnership capacity with Bosnia through focused security cooperation initiatives, to include International Military Education and Training, Foreign Military Financing, and the State Partnership Program with Maryland's Army National Guard.

In a show of its increasing capacity, Bosnia assumed a key leadership role during European Command's 2009 COMBINED ENDEAVOR exercise, involving 40 countries and 1,200 personnel. European Command remains focused on Bosnia's defense reform efforts and its entry into NATO, so that Bosnia can finally prosper, contribute more fully to coalition operations, and complete its path to Euro-Atlantic integration. Due to progress made, Bosnia was able to consistently contribute to the coalition efforts in Iraq between 2005 and 2008, and now has over 50 personnel deployed to ISAF. Despite some remaining challenges, we have achieved quite a lot given what was happening in Bosnia just 15 years ago.

Russia

The complexities of establishing and maintaining a military-to-military relationship with Russia are many. On one hand, there are many areas of potential cooperation and partnership, including Afghanistan, arms control, counterterrorism, counter-piracy, counter-narcotics, and missile defense. Continued open dialogue can lead to additional opportunities for cooperation and openness, such as European Command's response to assist Russia during last year's wildfires. On the other hand, some of our allies and friends in the region remain concerned about Russian actions, including the conflict in Georgia in the summer of 2008 and Russia's continuing suspension of implementation of the Conventional Armed Forces in Europe (CFE) Treaty.

Working with Russia is about balance and seeking to find the potential for cooperation, while maintaining an open and honest dialogue about all aspects of our relationship, including where we disagree. While a great deal of engagement with Russia is handled by either the State Department, in the diplomatic realm; or directly by the Joint Staff and the Office of the Secretary of Defense, we at European Command are ready to pursue military-to-military communication, engagement, and even joint training and operations with Russia, where and when appropriate.

In 2009, for example, European Command authored a framework document to resume military-to-military cooperation with Russia in an equal, pragmatic, transparent, and mutually beneficial manner. The framework not only addresses crisis response and consequence management operations, but also seeks to promote interaction and ensure mutual support in conducting counterterrorism and counter-piracy operations; peacekeeping missions; ballistic missile defense; and search and rescue.

This framework document, signed by the Chairman of the Joint Chiefs of Staff and the Russian Chief of Defense at the 2009 Presidential Summit in Moscow, began to rebuild a structure for our bilateral defense relationship that allows wide-ranging and candid engagement on all issues of concern. This effort is then supported by European Command's lead in developing the annual military-to-military work plan, which defines the events and activities that we aim to accomplish together over the next year.

Admiral James Stavridis, SACEUR, during an Honor Guard Ceremony in Moscow, Russia.

Russian amphibious ship, the RFS Kaliningrad, sailing alongside the USS Mount Whitney as parts of the BALTIC OPERATIONS exercises in the Baltic Sea.

In working the bilateral military-to-military relationship with Russia, however, European Command will work with NATO and other partners to implement an integrated and inclusive security cooperation architecture beneficial to all participants that does not come at the expense of our allies and partners.

Israel

The political/military environment in which Israel exists is volatile and uncertain. Israel faces frequently voiced threats from Hezbollah, Hamas, Iran, and others in the region. Iran's aggressiveness and negative rhetoric could also pose serious security challenges to the region broadly, to the United States, and to other allies.

In addition, conflicts such as the Lebanese conflict of 2006, the Gaza hostilities of 2008, or the cross-border shooting in August 2010 with Lebanon, could erupt at any moment. Conflict could result from instigation by Iran or its proxies, notably Hezbollah, or from miscalculation.

European Command's engagement with Israel continues to strengthen our relationship with this key regional ally. We conduct multiple Headquarters and Component-level theater security cooperation events annually with Israel and chair four bilateral, biennial conferences spanning planning, logistics, exercises, and interoperability.

The United States and Israel routinely conduct training exercises that build partnership and work toward regional stability. This exercise portfolio includes eight major recurring exercises.

The continued success of the exercise program improves interoperability, understanding, and cooperation between the Israeli Defense Force and U.S. military forces. European Command leadership and staff maintain uniquely strong, recurring, personal, and direct relationships with their counterparts in the Israel Defense Force.

Turkey

Turkey remains a strong ally and partner in the region and continues to grow in importance in the Middle East and Eurasia regions. Turkey will continue to play an important role in the fight against extremism, maintaining regional security and access, deterring common threats, and supporting NATO out-of-area operations, such as those in Afghanistan and Kosovo.

As our presence in Iraq draws down, Turkey's concern with possible volatility on their border may grow, driving them to play a larger role in regional stability. We have continued to develop new and productive ways to increase our engagement with Turkey in military-to-military areas, and as part of U.S. interagency efforts to help this important and centrally-located ally face the challenges posed by 21st century threats.

As part of that effort, Special Operations Command Europe continues to engage in a highly productive program of tactical exchanges and training events with Turkish Special Forces, and has focused part of this outreach program to develop, enhance, and convey lessons learned by U.S. Special Operations Forces over the past eight years in the effective fusion of intelligence, operations, and interagency coordination.

Also this year, European Command initiated Exercise COOPERATIVE RESOLVE, a new bilateral exercise between U.S. Army in Europe forces and Turkey Land Forces Command, focused at the battalion-level to share best practices and emerging concepts for command post operations, command and control functions, and other associated tactics, training, and procedures.

Ukrainian marine 1st and 2nd divisions stormed the beaches of Tenrda island from amphibious assault ships and helicopters to simulate landing on an island over run by pirates during a combined and joint maritime exercise in the Black Sea.

On another front, European Command's new Joint Interagency Counter-Trafficking Center (JICTC)-Europe, an information-sharing, internally-resourced collaborative enterprise, is working closely with our Turkish partners to assist and enhance the capabilities of Turkish agencies to combat illicit trafficking and terrorism, two often-interrelated and interdependent activities that threaten the security of our partners, our theater, and our homeland.

Finally, European Command continues its highly successful Turkish/U.S. officer exchange program in which each country selects a group of promising, young field-grade officers for travel and exposure to each other's military headquarters and facilities, building vital relationships on a personal level that will continue to pay dividends as these officers progress to higher ranks and greater levels of responsibility.

Terrorism in Europe

Our contribution to the ever-evolving fight against transnational and indigenous terrorism continues to be one focused on intelligence sharing and partner-building. Across Europe, the threat of terrorist attack and the presence of terrorist support networks remains a serious concern. Violent-minded extremists seek to exploit vulnerabilities to radicalize local populations and logistically support jihadist war-zones such as Iraq and Afghanistan. The European continent remains a priority target for directed attacks from alQaida and allied terrorist groups who believe that lethal attacks will weaken the resolve of our partner nations to continue supporting the International Security Assistance Force.

The terrorist attack in Stockholm on December 11, 2010 and the late-December arrests of terrorist suspects in the United Kingdom and Denmark, among others, are indicative of a continuing threat.

The threat from al-Qaida associated elements remains high, particularly in Germany, Denmark, France, the United Kingdom, and Belgium. Notably a recent study by the Danish Institute for International Studies found that between 2004 and 2008, 84% of terrorists detained in Europe had a Western upbringing, and only 28% had links to foreign militant groups. This is indicative of the rise of "home-grown", self-motivated terrorists who, despite their exposure to Western values, chose to learn jihad over the internet and act without direct guidance by alQaida senior leadership.

European countries continue to improve their capacity to counter the terrorist threat by strengthening counterterrorism legislation, expanding international cooperation, and successfully prosecuting and jailing terrorist actors. To support this progress, European Command increases intelligence sharing at every opportunity. For instance, one of our projects disseminates evidence and information obtained by coalition allies on the ground in Iraq and Afghanistan to International Criminal Police Organization (INTERPOL) member countries' police forces worldwide in order to strengthen their counterterrorism efforts. This project has assisted investigations in more than 70 countries.

Ballistic Missile Threat

There is an existing and expanding threat from ballistic missiles to the European Command's area of focus. The continued development of missile technologies by states such as Iran and Syria, coupled with the transfer of rocket and missile capabilities and technologies to non-state actors such as Hizbollah, present the most significant combination of capability and intent into realized threat to European Command's interests in Europe and the Levant. Iran in particular, with its growing inventory of ballistic missiles, views its conventionally armed missiles as an integral part of its strategy to deter, intimidate and retaliate against forces in the Middle East, Southeastern Europe, and Central Asia.

Combating Weapons of Mass Destruction (WMD)

Nuclear weapons in the hands of a terrorist or violent extremist would represent a grave threat to the United States and our allies. The threat of a deliberate attack with a biological weapon, or the spread of nuclear weapons programs are threats that have far-reaching, destabilizing consequences. Al-Qaida and other groups aspire to incorporate weapons of mass destruction into their attacks. Special nuclear materials and the majority of the world's nuclear weapons are located in the European Command area of focus.

To succeed in preventing the spread of weapons of mass destruction and their precursor materials, we must pursue a vigorous, comprehensive strategy sustained by a whole-of-government approach and strong international partnerships. The security of the weapons and these materials is a significant aspect of the Command's efforts to counter WMD. We must support partner and allied efforts to build capacity that detects and interdicts the movement of WMD materials, facilitates information-sharing arrangements, and, if an attack occurs, ensures a swift effective response that includes consequence management. Recent concerns over attacks in Europe truly underscore the importance of working with our partners and allies to prevent a catastrophic attack involving weapons of mass destruction on their soil and our homeland.

> "Whether it is ensuring our capabilities for strategic missile defense, which provides assurance to allies and deterrence to adversaries, or making the necessary investments in defending against cyber attacks; we must be able to look towards the future and see likely security challenges that we will face and meet them head on."
>
> Chairman Howard P. "Buck" McKeon, House Committee on Armed Services.

Cyber Security

Cyber security is a vital function for European Command, and we have made progress in securing our part of cyberspace during the past year. We have joined the efforts of other combatant commands, including the leadership of the U.S. Strategic Command and its sub-unified command, U.S. Cyber Command. The establishment of U.S. Cyber Command is a great step forward for the Department and all combatant commands. We are exploring the paths to more closely fuse our internal efforts in cyberspace, and have named a Cyber Integrator on our Headquarters Staff to build on our established Cyber Fusion Center. As the number and sophistication of attempts to penetrate our network increase daily, we must continue to meet that threat with innovation and expertise because we cannot afford more manpower.

As we gain operational experience in cyberspace, we look for opportunities to provide bridges between industry, academia, government, and our military partners with the goal of increasing mutual awareness and security. There are many challenges in this new domain and several of our friends throughout Europe have been the subject of cyber harassment (disruption of cyberspace functions) coincident with international disputes. As we look for answers and solutions to the evolving challenges in cyberspace, we synchronize frequently with our partners at various exercises and conferences, to include Exercise COMBINED ENDEAVOR 10, which brought together forty nations focused on enhancing common awareness, building common understanding, and developing operational trust, as well as a 2011 European Cyber Conference Symposium hosted at the George C. Marshall European Center for Security Studies. We have several other visits and exercises planned in 2011.

As we move forward, we are looking to leverage our existing European Union and NATO relationships in concert with the newly formed U.S. Cyber Command to increase our shared security. With NATO's desire to develop an infrastructure to meet the cyber security threat with a coordinated combined response, we are moving forward now to lay the groundwork for what we anticipate will be our role in this combined action.

The Arctic

The Arctic maritime domain is changing, and nations are responding by enacting policies to address anticipated challenges in the region. From environmental impacts to commercial enterprise, the prospect of unprecedented access to natural resources and Northern shipping routes has raised related security concerns. Promoting regional stability in the Arctic is European Command's primary objective for the Arctic. This is best accomplished through the use of open international forums to ensure stable, predictable management of the Arctic area.

In areas of safety, security, resource management, and conflict resolution, we support multinational governance and the international rule of law. Development of peaceful and successful maritime domain awareness is essential, not only to the security of the region, but also to the safety of the mariners, workers, and tourists who will populate this vast expanse. European Command is committed to a future that includes collaboration with our partners, NATO members, and Russia on international search and rescue exercises, Arctic training, and transparent operations and diplomacy that fully respect territorial claims.

Addressing environmental security in the Arctic region will require close cooperation with a wide range of Arctic stakeholders. Of particular note, we see Russia as a key potential partner in this area; one with substantial capabilities to respond to unforeseen emergencies and a clear willingness to protect the region from environmental disasters. We look forward to working with Russia and our other Arctic partners as we seek areas of mutual interest.

Our strategic national interests are served by fostering global economic development. A growing and open global economy enables the growth of the American economy, and that growth forms the backbone of our national security. Most of our major European partners are heavily dependent upon foreign energy sources for their oil and natural gas. This reliance can have a destabilizing effect on European economic development. Russia is one of the most important suppliers of crude oil and natural gas to Europe, accounting for 33% of oil imports and 40% of gas imports (87% for Italy; 81% for Spain; 61% for Germany; and 51% for France).

U.S. Submarine surfacing in the Arctic.

Energy Security

Russia's energy leverage represents a key factor in European and Eurasian energy security. Europe will continue to need Russian energy, as supplies from Russia are useful alternatives to reliance on the Persian Gulf for hydrocarbons and the pipeline infrastructure to transport it is already in place.

European Command supports State Department objectives regarding European energy security, and we work with our interagency partners, NATO allies, and partner nations to support these objectives: diversification of energy transportation routes in Europe; greater intra-Europe integration of existing supply systems; the development of new, renewable, and alternative energy sources in Eurasia; and demand-side efforts to promote energy efficiency. Within European Command, we are proposing a joint concept for energy security to achieve these same objectives, ensure access, and decrease vulnerabilities within our own forces. Our J9 Interagency Partnering Directorate continues to employ a whole-of-government approach to collaborate with our partners and like-minded allies to develop frameworks for addressing major energy security issues.

> "Our relationship with our European allies remains the cornerstone for U.S. engagement with the world and a catalyst for international action."
>
> National Strategy, May 2010.

INITIATIVES

Effective pursuit of U.S., allied, and partner interests depends ultimately on our ability to innovate and find new and better ways of achieving our objectives.

Support to NATO, Especially in Afghanistan

Today, European Command's largest contributions to support NATO lie in our efforts to train and equip partner nations to deploy alongside our own troops in Afghanistan. We have multiple lines of effort to support these activities.

Joint Multi-National Readiness Center

The Joint Multi-National Readiness Center supports European Command and Central Command operations by providing pre-deployment training to Europe-based U.S. forces and NATO Operational Mentoring and Liaison Teams slated for deployment to Afghanistan. Currently, the Center provides enduring observer/ controller support to the United States Security Coordinator Israel to train the Palestinian National Security Forces. Joint Multi-National Readiness Center observer/controllers were also instrumental in the successful pre-deployment training of the Jordanian 2nd Ranger Battalion for operations in support of Afghanistan's national elections. We have trained almost 4,000 soldiers to date and, through these efforts, European Command has enabled partner nations to make significant contributions to operations in Afghanistan.

Georgia Deployment Program-International Security Assistance Force

Marine Forces Europe directly supports the Republic of Georgia's two-year program to deploy Georgian forces alongside U.S. Marine Forces to Afghanistan. The Georgia Deployment Program-International Security Assistance Force will deploy four rotations of a Georgian battalion with a Marine Corps Marine Expeditionary Brigade to Afghanistan. As capabilities improve, Georgian forces will be able to operate independently. By using Georgian shadow instructors, Marine Forces Europe will create a Georgian training group that will largely take over the Partnership Training Program by their fourth rotation. Over this past year, this program trained two battalions that deployed to Afghanistan.

Interoperability

EUCOM has significantly enhanced the ability to communicate with NATO and NATO member nations at the SECRET level. The U.S. Battlefield Information Collection and Exploitation System (BICES) network is used for planning, exercising, and operating with our NATO partners in this theater. We exercised that capability in AUSTERE CHALLENGE 10, JACKAL STONE 10, and a NATO Cyber Defense Exercise. As we leverage these opportunities to hone our ability to work together, we will continue to determine the capability requirements and develop the tactics, techniques, and procedures to ensure that U.S. BICES meets our NATO interoperability needs.

Georgian soldiers from the 31st Battalion establish security during a cordon and search scenario as they train for an upcoming deployment in support of International Security Assistance Forces in Afghanistan.

In another area, the United States became a full participating member in the Military Engineering Center of Excellence. Through this body, our engineers have the opportunity to develop interoperability and relationships with engineer forces at all levels of command. We are able to provide expertise to other NATO countries and help them prepare for NATO operations. Through these engagements, our nation also benefits by learning new engineering methods, and gains access to the resources of the Center of Excellence.

Ballistic Missile Defense

The concentrated efforts by our adversaries to illicitly procure ballistic missile technology, develop increasingly sophisticated missiles, and actively refine their abilities to employ those missiles against friends and allies have not abated.

As we work to provide defenses for our deployed forces, families, friends and allies, European Command continues the extensive and active cooperation necessary to implement the European Phased Adaptive Approach (EPAA) to Missile Defense. Together with our partners in the Department of State, Department of Defense, Missile Defense Agency, and many others we are fully supporting the coordinated international engagement of the United States.

Our coordinated efforts are bearing fruit, with NATO declaring at the Lisbon Summit that it will develop a missile defense capability to provide full coverage and protection for all NATO European populations, territory, and forces. NATO reiterated its longstanding invitation for Russia to cooperate with the Alliance in this endeavor. We are also working with Poland and Romania who have agreed to host elements of our missile defense systems.

A Standard Missile - 3 (SM-3) is launched from the USS Hopper (DDG 70) in a Missile Defense Agency (MDA) test in conjunction with the U.S. Navy.

As part of these efforts, European Command is working with both the Missile Defense Agency and the Commander, Naval Installations Command, to ensure that facility infrastructure will be ready to go when the system is activated.

In order to provide for Communication, Collaboration, Coordination and, potentially, Command and Control of U.S. and NATO Missile Defense forces, European Command's J6 directorate has aggressively deployed U.S. BICES (NATO SECRET) workstations throughout European Command's Headquarters and our Service components. These workstations are supported by theater collaboration services for secure voice, chat, and information-sharing as well as the Battle Command Systems—providing U.S. and NATO forces with all the tools necessary to execute this mission successfully. This spring, European Command will add U.S. ships afloat to the U.S. BICES architecture, further integrating our sensors, shooters and platforms within theater.

At the same time, European Command is preparing for the arrival of the initial assets that will operationalize the European Phased Adaptive Approach. As we work together with our partners and allies to field an operational capability this year, I maintain my firm belief that the capabilities delivered will serve as a catalyst to develop a cooperative solution with our allies and partners in the region.

Joint Interagency Counter Trafficking Center

This past August, European Command stood up the Joint Interagency Counter Trafficking Center (JICTC) to support interagency efforts across the theater to counter the growing transnational trafficking threat. Transnational trafficking is a multi-faceted U.S. national security concern which has potential to undermine U.S. and international efforts to protect public health and ensure regional security.

Transnational organized criminal activity contributes to weakening the rule of law, and fosters other forms of illicit activity such as terrorism, insurgency, organized crime, weapons trafficking, money laundering, human trafficking and piracy. Left unchecked, this activity can continue to spread and metastasize, threatening the stability and legitimacy of key states, as well as the U.S. homeland.

European Command's trafficking center will complement the State Department's interagency programs and assist the international community and European national efforts to build self-sufficient border management skills, competencies, and capacity among partner nations.

The vision is that we will stand up a truly international, interagency organization focused on counter-trafficking. Though loosely modeled on Joint Interagency Task Force (JIATF)-South in Key West, our trafficking center will have important differences. Unlike JIATF-South, the Joint Interagency Counter-Trafficking Center will not directly participate in detection, monitoring, and interdiction operations on land or at sea. Instead, it will provide depth and capacity to our interagency partners and, by doing so, will provide another layer to regional security and the defense of our homeland. European

Command is also leveraging opportunities to link U.S. government agencies through information systems by building a common, user-friendly information technology portfolio that facilitates information- sharing and cross-cooperation.

We are developing agreements on information-sharing standards with partner military, police, and civilian organizations to support counter-trafficking and exploitation activities.

Our main focus will initially be on counter-narcotics, but will ultimately integrate other aspects of the transnational trafficking threat. No additional resources are needed to stand up the JICTC, and we plan to be fully operational by September 2011.

Whole-of-Government/Whole-of-Society Approach

Interagency partnering remains the heart of the enterprise for this Command, and is critical to how we approach security challenges in our theater. Building on the interagency cooperation that presently exists at our U.S. Embassy Country Teams, we have also grown the interagency presence at European Command to best effect interagency collaboration at the regional/operational level.

Since I last addressed you, we have welcomed additional representatives to European Command Headquarters from U.S. Customs and Border Protection, the Department of Energy, and the Department of Justice's Drug Enforcement Administration who wish to coordinate their activities with the U.S. military. This is in addition to representatives already in place at the headquarters from the Department of State, Department of the Treasury, Federal Bureau of Investigation, and Immigration and Customs Enforcement. Additionally, we are actively working with the U.S. Agency for International Development to once again secure their representation at European Command. Finally, the Department of Justice will soon add their prosecutorial and rule of law expertise to our team as, together, we seek to build partner capacity in Europe and Eurasia and, in so doing, better defend our homeland forward.

We have also introduced the U.S. Department of State Coordinator for Reconstruction and Stabilization (S/CRS) into our COMBINED ENDEAVOR exercise. This infusion of interagency participation facilitates cross-cooperation, coordination, and information sharing between the Departments of Defense and State in the context of whole-of-government operations within the European theater. This reinforces the Command's commitment to building stronger partner capacity, not only among our internal U.S. government agencies, but also with NATO and European partner government agencies as well. We look to integrate with both U.S. interagency organizations and those of NATO and our partner nations by coupling the U.S. whole-of-government approach with the comprehensive approach functions of our partners.

We have also established a J9 Interagency Partnering Directorate, the first new directorate at European Command since 1967. In addition to working with interagency partners, it also engages and collaborates with international and non-governmental organizations, academia, the private sector, think tanks, and military academic organizations. In the private sector, for example, we have gained many new insights by partnering with numerous organizations. As part of our efforts to engage leading European professionals, we have instituted the European Partnership Program to both listen to European business professionals and leaders and to speak directly to them about our mission in Europe and Eurasia. In addition, our new Academic Outreach function has leveraged the knowledge and fresh thinking of military and civilian academic institutions to help us in our military planning processes.

Public Private Cooperation

European Command has begun work to leverage the expertise and other assets of the private sector to achieve its objectives as well as support the efforts of NATO Allied Command Operations. We are currently working with private businesses as well as non-profit non-governmental entities. For example, in collaboration with Central Command, we organized a public-private workshop on further development of the Northern Distribution Network and the potential for building a "silk road," or commercial transportation network for commerce between Europe and Eurasia through Afghanistan. The participants included officials from the U.S. Transportation Command, Defense Logistics Agency, U.S. Central Command, U.S. Agency for International Development, the Asian Development Bank, the International Road Union, the National Defense Transportation Association, and other business executives, who agreed on recommendations for further action ondeveloping and implementing a silk road strategy as a component of the U.S. transition strategy for Afghanistan.

We are also looking at ways to incorporate this kind of collaboration for issues like cyber security, assessments, and humanitarian assistance.

Humanitarian Assistance Programs

European Command's Humanitarian Assistance programs directly benefit the nations where they are executed, and consist of: the Humanitarian and Civic Assistance Program; the Humanitarian Assistance-Other Program; and the Humanitarian Assistance Program-Excess Property. Projects funded through these resources complement U.S. Agency for International Development efforts, enhance regional security cooperation, and advance U.S. interests throughout the region. They also bolster a country's own capability to respond to disasters, thereby diminishing the need for future U.S. involvement, and provide an example of the value of a military in times of peace.

Interagency, academic, public, and private partners of the U.S. European Command.

While the European Command Humanitarian Assistance budget is relatively small, it has a disproportionately high and positive impact. Last year, the command executed $17 million in Humanitarian Assistance Project funding for 145 security assistance related projects across 18 countries.

One example of a Humanitarian Assistance project that was carried out in an interagency and public-private manner to gain efficiencies and maximize impact was our contribution to a nursing school in Georgia. We contributed through training to the construction of the facility, while Emory University staffed it and the U.S. Agency for International Development equipped it.

Another example was European Command's partnership with the U.S. Agency for International Development, an engaged non-governmental organization, and the Albanian Ministry of Health in an integrated effort to establish telemedicine capabilities throughoutAlbania.

This past December, six of 14 telemedicine centers were connected to enhance basic health care, disaster management and emergency response across the country. While European Command training assisted in renovation of the facilities, the impact of the combined effort provides a higher level of health care and continued medical education nationwide.

> "Innovation distinguishes between a leader and a follower."
>
> Steve Jobs, Co-founder and CEO of Apple.

Innovation

Implementing new ideas and innovations is vital to achieving success in today's complex and adaptive security environment. Indeed, success may well be determined in our labs, think tanks, and centers of innovation. The original discovery, development, and rapid implementation of technology and ideas are imperative for staying ahead of our adversaries who are continuously adapting and innovating as well.

Acting as a catalyst and accelerant, and working in close collaboration with the Office of the Secretary of Defense, the Services, the U.S. Interagency, industry, and our partner nations, our Innovation Cell has achieved tangible results in furthering unique and innovative technologies.

Results include discovery and enhancement of a unique human detection technology, a cross-domain solution for collaboration with our allies, and acceleration of a system to support space-based wireless internet access to remote regions.

Each of these successes has been achieved by partnering with and leveraging the unique technologies of our allies.

The Command has enabled access and connected unique partner nation technologies to counter improvised explosive devices, piracy, and smuggling. Future innovative projects planned are in the realm of information technologies to support population-centric counterinsurgency tools.

The Rosetta Stone and books that U.S. European Command has featured in its Command reading list, a project designed to enhance cultural understanding.

Cultural Understanding and Language Study

I have often talked about the power that comes from understanding a country's culture. At European Command, we are always trying to find ways to increase our understanding of European culture throughout our organization as we continue the important work of building and strengthening our relationships with partner nations. One example is our Next Generation Advisory Panel, a body of up-and-coming civilian leaders from several European countries with whom we engage on a regular basis to seek their unique perspective on a variety of issues. Another is our Notable Author Series, which brings prolific writers and thinkers to the Command whose books add important historical context into an open forum for engagement and discussion with our staff. Our Academic Outreach Division recently brought an expert on Russian and Central European security issues to the Command as part of the European Command Forum for International Affairs Speaker Series. Additionally, European Command's Strategic Languages Program has joined our headquarters in Stuttgart with the Defense Language Institute in order to provide foreign language training for our staff members.

Our Most Important Resource

None of the activities described in these pages would be possible without the people that make up U.S. European Command. We have a responsibility to our people and their families to ensure the readiness and health of our force and provide quality of life support to the families. Being stationed overseas presents unique challenges and opportunities for our force. Maintaining a balanced and efficient overseas force posture, however, is critical to the defense of the United States.

Deployment, Behavioral Health, and Compassionate Fatigue and Family Support

Protracted combat operations and multiple deployments have, as you know, placed significant stress on our service members and their families. Several organizations and studies within the Department have identified an urgent need for sustained behavioral health services to support these warriors and their family members. As we continue to maintain mission readiness, it is imperative that our dedicated military men and women, and their families, have access to these vital programs and services without stovepipes in a stigma-free environment. A 360-degree review of these programs, with an analysis of the connection between at-risk indicators and catalysts, is needed to eliminate gaps in support. The goal is alignment of focused caregiver teams with corresponding indicator data systems to ensure the health of our force and family. We will also continue to work with the Office of the Secretary of Defense to develop a working definition for resilience, while determining initial measures for baseline assessments to address at-risk indicators.

Additionally, we must also care for our vital caregiver teams. Last year, I testified on the state of European Command's community caregivers, who themselves have shown signs of stress, burn-out, and compassion fatigue. At that time, I briefed you on our comprehensive compassion fatigue program, entitled "Providing Outreach While Enhancing Readiness—Caring for the Caregiver," which focuses on providing caregivers with tools and strategies to prevent and mitigate the risk of stress, burn-out, and compassion fatigue. 2010 was the first year of execution for this initiative; one we plan to continue and one that is showing promising results. Lastly, we continue to support ongoing efforts to improve complex care management and the medical portion of the disability evaluation process, which will result in improvement of wounded, ill, and injured warrior benefits.

Theater Infrastructure

Thanks to strong and continued Congressional support, previous annual military construction authorizations and appropriations have enabled European Command to address a balanced mix of our most pressing mission, mission support, quality of life, and housing requirements. The goal of our fiscal year 2012 military construction program is to support ongoing force posture initiatives, consolidation efforts, and infrastructure recapitalization projects including the Kaiserslautern Military Community (KMC) Medical Facilities Recapitalization and Consolidation project. As always, when there are opportunities to leverage NATO common funded investments, we do so. And where required, we pre-finance our projects to reserve a future opportunity to recapture a portion of our investments through the NATO Security Investment Program. At enduring locations, we continue to sustain and recapitalize our infrastructure through responsible use of both the Sustainment, Restoration, and Modernization program and the military construction program.

At non-enduring locations, we are optimizing use of all available resources to ensure these installations remain fully mission effective until the installations are removed from the inventory. To that end, European Command's footprint currently includes approximately 350 distinct real estate sites (ranging in size from small unmanned communication sites to Ramstein Air Base), which collectively make up the present Command footprint, down from

1,200 during the Cold War. Anticipated changes, some of which are planned within enduring installations, will result in the return of approximately 100 of these sites to host nations soon. We are constantly reviewing requirements across the current and new mission, quality of life, and agency portfolios to work towards joint solutions where appropriate. European Command's future requirements will appear in our Theater Posture Plan and military construction requests. The sites are all tax and rent free and receive much host nation support.

We are pleased that the Department of Defense Education Activity (DoDEA) is planning and executing a $2.1 billion investment into DoDEA's Europe school infrastructure that has been in need of support for many years. Many of our schools are converted barracks from the 1950's and will benefit tremendously from this investment. At the end of this seven-year program, all failed or failing infrastructure will be recapitalized, providing concrete proof of our priority to take care of our people and their families. Additionally, we will continue to address and pursue improvements to our military family housing and barracks/dormitories in the 2012 military construction program to improve the living conditions of our families.

Force Posture

The presence of U.S. forces in the European theater serves many important functions. It fosters relationships and deepens partnerships with individual countries as well as an entire region of significant importance to U.S. global strategic interests, as evidenced by the overwhelming number of ISAF troop-contributing nations that come from this theater.

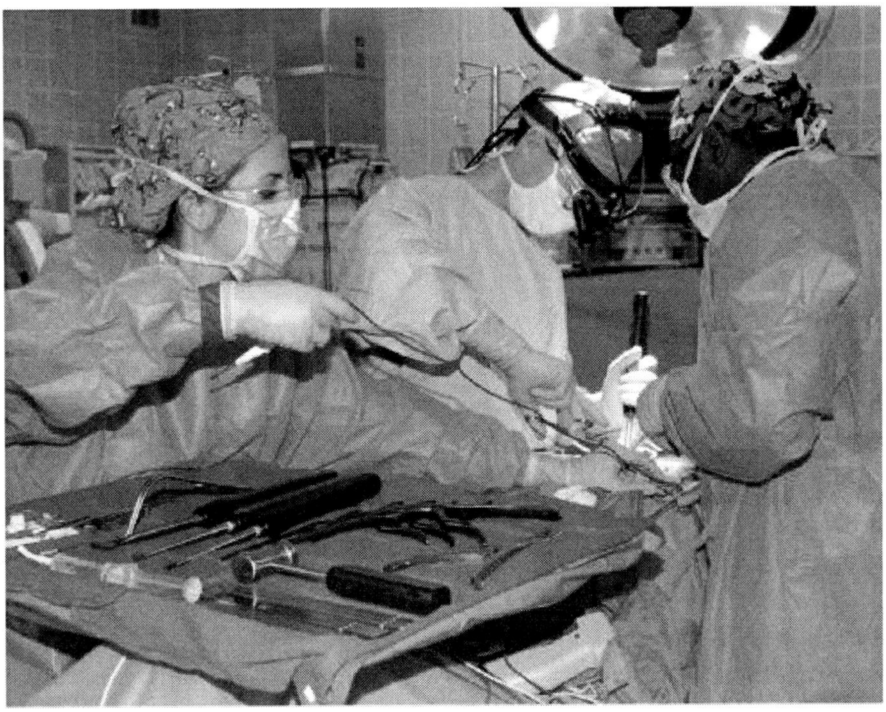

A surgical team from Landstuhl Regional Medical Center, Germany, operates on a Soldier injured in combat operations.

This continuous presence and partnership allows European Command to train alongside our allies and partners to build their capacity as well as our own, and increase interoperability. U.S. forces stationed in Europe today act to assure our allies even as they deter and dissuade our adversaries, and are the most visible indication of the ongoing U.S. commitment to the NATO Alliance. European Command's footprint also enables the projection of U.S. power globally. Sites and installations in Europe provide superb power projection facilities for the support of coalition operations and overseas contingency operations.

As we consider U.S. presence overseas, we must consider the security environment in which we are currently operating. As the post-Cold War security environment changed, the size of our forces saw a corresponding change. The number of active-duty U.S. personnel in Europe has gone from over 400,000 during the Cold War to approximately 80,000 today. With ongoing activities in Afghanistan and Iraq, we assess that we should maintain our asset levels to maintain our current levels of effort in the immediate future, to include deployment rotations and partner training schedules. As our engagement requirements change, we will also look to adjust our asset levels. As the Secretary of Defense has said, "Based on our review, it is clear we have excess force structure in Europe. We are looking closely at alternative courses of action, but none would be implemented before 2015 or without consulting our NATO allies." In doing so, we not only look at pure numbers of troops when examining force posture, but also at capabilities and force mix. As the mission in Afghanistan begins to draw down, we may begin to reduce in the area of combat troops. However, as our ballistic missile defense mission develops in the near future, we will also evaluate the force posture needs associated with that growing mission.

There are five significant force posture initiatives that European Command is undertaking to support building the capability and capacity of partner nations in Europe, increase expeditionary capability from Europe, support other combatant commands, and achieve basing efficiencies.

The first initiative is in direct support of the EPAA. The EPAA is the U.S. voluntary national contribution to NATO missile defense. This will be an integral component of NATO's mission to provide full protection and coverage for all NATO European populations, territory, and forces as well as enhance the defense of the U.S. homeland.

The second initiative is an Unmanned Aerial Systems Center of Excellence at Naval Air Station Sigonella, Italy. The synergistic impact of combining U.S. Air Force Global Hawks, U.S. Navy Broad Area Maritime Surveillance, and NATO Alliance Ground Surveillance unmanned aerial systems programs at one location within close proximity to three geographic combatant commands is a prime example of how the European Command is maximizing our efficiency within the European Theater.

The third initiative is the timely stationing of the Joint Strike Fighter into theater. The proper timing of the Joint Strike Fighter bed-down ensures that the U.S. maintains its leadership role within the NATO Alliance.

The fourth initiative is developing a U.S. Transportation Command requirement for a Black Sea/Caucasus en-route location to further U.S. expeditionary capability. The European Command will meet this requirement while maximizing our basing efficiencies.

The final initiative, is providing direct support to U.S. Africa Command, which is located in Europe. The European Command, because of our global strategic location, is properly positioned for other combatant commands and interagency partners to leverage our resources. We consider it our responsibility to maximize efficiency in the theater.

NATO/SHAPE

> "Europe and North America have known more than six decades of peace and stability. But this has not come for free. It has required political vision and courage. It has required an enormous, sustained investment – in both political and financial terms – to defend not just our territories, but also our values. And it has required – above all – strong transatlantic solidarity."
>
> NATO Secretary General Anders Fogh Rasmussen, Speech given at NATO Lisbon Summit, November 2010.

NATO has been the anchor of Trans-Atlantic security for more than 60 years, ensuring the security of its members, enhancing peace and stability throughout Europe, and countering threats across the globe. In November of 2010, the Heads of State and Government of the Alliance approved a new NATO Strategic Concept at the Lisbon Summit and mandated a series of actions to modernize and enhance the Alliance's capability to address the complex challenges of this era. The Summit was a pivotal event in the Alliance's history, framing its future and demonstrating the political will of its members to strengthen our individual and collective readiness and capabilities for the full range of security challenges.

New Strategic Concept

NATO's new Strategic Concept, the first in ten years, is titled "Active Engagement, Modern Defense". The Strategic Concept reconfirmed the bond between all members to defend one another against attack, including against new threats to the safety of our populations. It committed the Alliance to prevent crises, manage conflicts, and stabilize post-conflict situations, including by working more closely with our international partners, most importantly the United Nations and the European Union (EU).

It offers NATO's partners around the globe more political engagement with the Alliance, and a substantial role in shaping the NATO-led operations to which they contribute. It restates the Alliance's firm commitment to keep the door to NATO open to all European democracies that meet the standards of membership, because enlargement contributes to the goal of a Europe whole, free and at peace.

The Strategic Concept also commits NATO to continuous reform towards a more effective, efficient, and flexible Alliance. The Strategic Concept reaffirms the fundamental purpose of the NATO Alliance and defines three core tasks for the Alliance: Collective Defense; Security through Crisis Management; and Cooperative Security through Partnership.

Collective Defense

On Collective Defense, the new Strategic Concept reaffirms the Alliance's core mission of mutual defense as set out in Article V of the Washington Treaty. This Article V commitment remains firm and binding. NATO will deter and defend against any threat of aggression, and against emerging security challenges where they threaten the fundamental security of individual Allies or the Alliance as a whole. The Strategic Concept mandates the maintenance of Alliance Deterrence, based on an appropriate mix of nuclear and conventional capabilities, and reaffirms deterrence as a core element of our overall strategy. The Alliance also agreed to pursue missile defense as a core element of Alliance defense and deterrence.

Crisis Management

Concerning Crisis Management, the new Strategic Concept commits the Alliance's unique and robust set of political and military capabilities to address the full spectrum of crises—before, during, and after conflicts. It recognizes that crises and conflicts beyond NATO's borders can pose a direct threat to the security of Alliance territory and populations. The Strategic Concept and the Lisbon Summit Declaration highlight the importance of a "Comprehensive Approach to Crisis Management."

NATO will engage, where possible and when necessary, to prevent crises, manage crises, stabilize post-conflict situations and support reconstruction. NATO will actively employ an appropriate mix of political and military tools to help manage developing crises that have the potential to affect Alliance security before they escalate into conflicts, to stop ongoing conflicts where they affect Alliance security, and to help consolidate stability in post-conflict situations where that contributes to Euro-Atlantic security.

A modernized and comprehensive approach to crisis management will involve engaging actively with other international actors before, during, and after crises to encourage collaborative analysis, planning, and conduct of crisis management activities. It also requires a capability to monitor and analyze the international environment to anticipate crises and, where appropriate, take active steps to prevent them from becoming larger conflicts.

The role accorded to Crisis Management in the Strategic Concept also reaffirms NATO's unique and essential role as a transatlantic forum for consultations on all matters that affect the territorial integrity, political independence, and security of its members as set out in Article IV of the Washington Treaty.

NATO Secretary General Anders Fogh Rasmussen unveiling the new Strategic Concept at the NATO Lisbon Summit.

Cooperative Security and Partnership

With respect to Cooperative Security and Partnership, NATO's new strategic concept recognizes that the Alliance is affected by, and can affect, political and security developments beyond its borders.

This concept mandates NATO to engage actively to enhance international security: through partnership with relevant countries and other international organizations; by contributing actively to arms control, non-proliferation and disarmament; and by keeping the door to membership in the Alliance open to all European democracies that meet NATO's standards.

In particular, the Strategic Concept highlights the importance of enhancing collaboration with the European Union and United Nations. It also stresses the priority accorded to forging a true, strategic partnership with Russia. In particular, the Alliance will pursue cooperation with Russia in the sphere of Missile Defense as well as enhance our cooperation in counter-piracy, counter-narcotics, counter-terrorism, and ongoing International Security Assistance Force operations.

In Lisbon, the NATO Heads of State and Government Summit Declaration contained many taskings related to implementing, or operationalizing, the new Strategic Concept. In particular, Crisis Management, Comprehensive Approach, Partnership, and Missile Defense will be focus areas for NATO Headquarters, Allied Command Operations, and Allied Command Transformation. The Lisbon Summit also tasked continued reforms for NATO in many spheres related to reducing costs and delivering efficiencies and effectiveness. NATO will continue to implement these important Lisbon decisions simultaneously and we will continue to conduct operations of high importance to our collective security.

> "All of the things that we've been working on for the last two years came to fruition in an extraordinary way in Lisbon,"
>
> *Secretary of Defense Robert Gates, November 2010.*

The USS Dwight D. Eisenhower Carrier Strike Group crew spells out NATO on the deck.

NATO-European Union Relations

The European Union is another potential partner for NATO in its Comprehensive Approach and, although slight, some progress has been made in the areas of cooperation and coordination between these two entities. In land operations, tactical coordination continues and, in Afghanistan in particular, there has been a growing willingness on both sides (International Security Assistance Force and European Union Police Mission Afghanistan) to coordinate efforts. This will hopefully lead to a more complementary approach, combining resources and capabilities to build Afghan National Security Forces' capacity. In the fight against piracy, NATO and the European Union have agreed to share tactical information for increased situational awareness and synergy. There have been other examples of tactical cooperation such as a recent incident during which an EU ship refuelled a NATO ship at sea. This common use of logistics support is an area that offers potential for further cooperation between the European Union and NATO.

Afghanistan

NATO's operation in Afghanistan continues to provide the Alliance a catalyst for change to ensure timely and relevant support to our combat forces. America's Allies in NATO have shared the risks, costs, and burdens of this mission from the beginning. They have contributed to the International Security Assistance Force and the Afghan National Security Forces and have made significant non-military contributions as well.

Afghan National Police and Afghan National Army personnel receive technical training on night vision goggles and thermal cameras by a Turkish officer.

The situation in Afghanistan today is complicated and deeply challenging, as externalpressures are balanced with internal recovery from 30 years of warfare. As we proceed in this campaign, the successful transition of security responsibilities remains the key issue. Much has been achieved in the past 12-18 months. The troop surge of 30,000 U.S. and 10,000 allied troops has had a significant impact on the ground, especially in southern Afghanistan. More importantly, the restructuring of the headquarters, including the activation of the ISAF Joint Command and NATO Training Mission-Afghanistan, have provided the ISAF Commander the leadership capacity to implement a counterinsurgency campaign focused on securing the Afghan population, developing the Afghan National Security Forces, and engaging the Afghan Government as a catalyst of change. We have largely halted the expansion of the insurgency, and are beginning to show signs of progress toward Afghan security self-reliance. These strategy reviews and increased attention on Afghanistan are welcomed by our allies and partners as we move forward. Our allies have already contributed a great deal to this war, fighting, bleeding, and dying side-by-side with our own troops.

There are four areas in which we must succeed in order to win in Afghanistan, and some progress has been made across all four. The first is to achieve synergy between our civilian and military efforts. To help accomplish this, NATO has reinforced the Senior Civilian Representative position with Ambassador Mark Sedwill. His efforts, in parallel with the United Nations Assistance Mission in Afghanistan, have shown exceptional progress in governance and development. Ambassador Sedwill and his team are providing the necessary balance to the military work being done by General David Petraeus, the Commander of U.S. and NATO forces in Afghanistan. The Senior Civilian Representative's efforts cannot be taken in isolation. Additional civilian expertise is still required to mentor, coach, and guide the Afghan government to take active visible steps to show that it is stamping out corruption, improving efficiency, and delivering necessary services to its people effectively.

The second area critical to the Afghanistan counterinsurgency effort is that the Afghans themselves must be at the center of this effort. In this aspect, the restructuring of the headquarters and the Senior Civilian Representative have provided the leadership capacity to engage appropriate Afghan leaders and move towards independence. The Afghan people, through village and district elders and shuras, have begun to assume responsibility for the wellbeing of their country and are showing growing confidence in their own government's ability to provide basic security and services without corruption and tribal favoritism.

A third important key to success in Afghanistan is effective strategic communication. A continuous flow of information that serves to bolster our actions is essential to assuring the Afghans, as well as our enemies, that the United States and our allies are committed to a secure and stable Afghanistan. We have work to do, but are improving.

Fourth and finally, as has been reaffirmed time after time by the Alliance, the most important role that the military can play in Afghanistan is to increase the size and capability of the Afghan National Security Forces (ANSF) through training and mentoring so that they may be able to take lead responsibility for securing their country. This is—and remains—the top resourcing priority in Afghanistan. Although the progress of NATO Training Mission – Afghanistan (NTM-A) and the increase in capacity and capability of the ANSF has been described as miraculous, trainers and mentors are still needed. Progress has been exceptional. And watching the enthusiasm as record numbers of recruits train at the Kabul Military Training Center definitely perpetuates optimism.

Recent polls have shown positive indications that progress is being made in Afghanistan. Almost 60% of Afghans believe their country is heading in the right direction. [1]Afghans are 83% confident that the Afghan National Army can provide security in their area and 75% confident in the Afghan National Police[2] . Thousands of insurgents are being captured or killed and hundreds of improvised explosive devices have been recovered. These are all indicators that validate our effort to put the Afghan people at the center of the equation in Afghanistan. We need to continue giving the Afghan people hope that they are not destined to live under the yoke of tyranny, and offer them every opportunity to live in a future Afghanistan worthy of their sacrifices.

Kosovo

Today, approximately 8,000 troops, including 800 MS soldiers, from NATO's Kosovo Force are deployed in Kosovo, working alongside local authorities to increase self-reliance in a multi-ethnic environment. The Allies decision to continue force reductions while developing internal security forces is the best declaration of this safe and secure environment. Operations remains challenging as tensions have potential to flare quickly but, by March 2011, planned force levels will be reduced to about 5,000 as five sites of historical and political importance have already transferred from Allied security to local authorities without incident. Following Kosovo's declaration of independence in February 2008, the Alliance reaffirmed that the Kosovo Force shall remain in Kosovo on the basis of United Nations Security Council Resolution 1244. NATO and the Kosovo Force will continue to work with the authorities and assist the United Nations, European Union, and other international actors, as appropriate, to support the further development of a stable, democratic, multi-ethnic, and peaceful Kosovo.

NATO and Iraq

At the Istanbul Summit in June 2004, the Allies agreed to be part of the international effort to help Iraq establish effective and accountable security forces. The outcome was the creation of the NATO Training Mission in Iraq (NTM-I), which to date has trained over 14,000 Iraqi security sector personnel.

NTM-I is involved in police training, establishing and mentoring Iraq's military academies, and facilitating substantial equipment donations and regular out-of-country training hosted by NATO Allies. All NATO Allies contribute to the training effort through deployment of trainers, provision of equipment, or NATO's financial contribution. The Government of Iraq regularly praises NTM-I, and continues to request its continuation and expansion.

ACTIVE ENDEAVOR

Under Operation ACTIVE ENDEAVOR, NATO ships are patrolling the Mediterranean and monitoring shipping to help detect, deter, and protect against terrorist activity. The operation evolved from NATO's immediate response to the terrorist attacks against the United States on September 11, 2001, and, in view of its success, is continuing. As the Alliance has refined its counter-terrorism role in the intervening years, the experience that NATO has accrued in ACTIVE ENDEAVOR has given the Alliance unparalleled expertise in the deterrence of maritime terrorist activity in the Mediterranean Sea. NATO forces have hailed over 100,000 merchant vessels and boarded 155 suspect ships.

By conducting these maritime operations against terrorist activity, NATO's presence in these waters has benefited all shipping through the Straits of Gibraltar. Moreover, this operation is also enabling NATO to strengthen its relations with partner countries, especially those participating in the Alliance's Mediterranean Dialogue.

Supporting the African Union

Well beyond the Euro-Atlantic region, the Alliance continues to support the African Union in its peacekeeping missions on the African continent. Since June 2007, NATO has assisted the African Union Mission in Somalia (AMISOM) by providing airlift support for African Union peacekeepers.

Following renewed African Union requests, the North Atlantic Council has agreed to extend its support by periods of six months on several occasions. NATO also continues to work with the African Union in identifying further areas where NATO could support the African Standby Force.

NATO's continuing support to the African Union is a testament to the Alliance's commitment to building partnerships and supporting peacekeeping and humanitarian efforts beyond the Euro-Atlantic region.

OCEAN SHIELD

Building on previous counter-piracy missions conducted by NATO beginning in 2008 to protect World Food Program deliveries, Operation OCEAN SHIELD is focusing on at-sea counter-piracy operations off the Horn of Africa.

Approved in August 2009 by the North Atlantic Council, the current operation, working with almost 40 ships from allies and partners in the context of the Contact Group on Piracy off the Coast of Somalia, continues to contribute to international efforts to combat area piracy.

This operation challenges normal paradigms, with information-sharing and coordination as the keys to success. These operating forces, from four different task forces under different mandates, have had an impact coordinating efforts through NATO's shared awareness and de-confliction efforts.

These efforts, along with the commercial shipping industry's strong encouragement of best management practices, have forced changes in the way the pirates operate; they have adapted by moving farther out into the Indian Ocean, and we must adapt accordingly. Although piracy inthe Gulf of Aden has been somewhat reduced, the overall number of hijackings has increased, as have the number of hostages held by the pirates.

Danish warship, HDMS Esbern Snare, stops and boards a suspected pirate vessel in the Gulf of Aden as part of Operation OCEAN SHIELD.

It is clear that, a longer-term strategy to build regional counter-piracy capacity is required, including clarification in international law of jurisdiction for pirates apprehended in international waters, as well as responsibility for their trial and incarceration if found guilty. This is under discussion among the allies.

NATO Special Operations Forces

The U.S.-led NATO Special Operations Forces (SOF) Coordination Centre was officially rechristened and activated as the NATO Special Operations Headquarters in November 2010.

The NATO Special Operations Headquarters, projected to be fully operational in 2012, has already had a significant impact coordinating, supporting, training, and enabling functions for NATO SOF, and it continues to develop Alliance crisis response options.

The evolution of this headquarters will better synchronize special operations forces across the Alliance, enhance NATO SOF unity of effort, and provide Allied special operation forces with a multi-national out-of-area command and control capability.

The NATO Special Operations Headquarters Communications Network underpins allied and partner SOF collaboration by providing an unprecedented vehicle for command, control, communications, and intelligence-sharing for networked operations. The Headquarters' Special Operations Forces Fusion Cell, in Kabul, Afghanistan, demonstrates this operational impact among allied and partner special operations forces. This stakeholder-run enterprise, manned by 40 personnel from 11 nations and several agencies, focuses on garnering information from a multitude of allied and partner sources, and fusing that information with operational requirements to produce and disseminate actionable intelligence to ISAF SOF Special Operations Task Groups and our Afghan partners.

NATO SOF Forces conduct nighttime training.

The NATO Special Operations Headquarters is building enduring operational capabilities, collaborative policies and procedures, and networked command, control, and communications mechanisms among NATO special operations forces. Collaborative training and exercises reinforce this framework to ensure allied and partner special operations forces are interoperable in order to operate more effectively in designated combined operations well into the future.

NATO Non-Commissioned Officer Initiatives

The first ever NATO Non-Commissioned Officer Bilateral Strategic Command Strategy and Recommended Non-Commissioned Officer Guidelines was published in October 2010. This first examination of Alliance-wide Non-Commissioned Officer Corps utilization defines critical gaps based on listed assumptions and implications, and delivers recommendations on how best to address them. It also outlines desired leadership qualities required in a multi-national environment, addresses NATO noncommissioned officer education, and explores the use of Command Senior Enlisted Leaders to assist the Commander in the professional development of the Non-Commissioned Officer Corps in order to better meet the demands of working in the NATO Alliance. Additionally, our Command's Senior Enlisted Leaders have led NATO efforts to assist partner nations with Non-Commissioned Officer reform in several countries this year through their involvement with initial assessments. We are working to tie these NATO initiatives into European Command's theater campaign plan to assist our partners with their Non-Commissioned Officer transformation. We believe these efforts will result in the more effective use of the Non-Commissioned Officer Corps—an essential component to achieving success in a multi-national environment.

CONCLUSION

"This reflects an enduring truth of American foreign policy – our relationship with our European allies and partners is the cornerstone of our engagement with the world, and a catalyst for global cooperation...Neither Europe nor the United States can confront the challenges of our time without the other."

President Barack Obama, New York Times, November 2010.

The Soldiers, Sailors, Airmen, Marines, and Civilians of European Command and Allied Command Operations contribute to our national security everyday through their professional engagement with our allies and partners across the European theater. As we look forward to continued success, I ask for your continued support of these extraordinary men and women and their families to ensure they receive the care and benefits they deserve.

Operationally, we will continue to seek and use flexible authorities and funding mechanisms to build the capacity of those partner nations willing to contribute to current operations. This has become increasingly important because of the surge of activities in Afghanistan and the need to get our allies and partners more involved. Your continued support for authorities like NDAA Section 1206, Foreign Military Financing, the International Military Education and Training program, and Coalition Readiness Support Program has been pivotal in addressing our strategic needs in the European theater, not only for partner-nation forces deploying to Afghanistan, but for all of our other allies to help build partner capacity. These programs allow us to provide them with equipment and training necessary to achieve interoperability with our own forces, and better prepare them to handle the responsibilities to which they commit their forces. Furthermore, our efforts to fulfill this short-term task of building enduring capability are vital to ensuring the long-term stability and security of Europe. In addition to increasing the contributions of our allies and partners to operations outside Europe, building partnercapacity allows us to make significant progress toward achieving European Command's strategic objectives. For example, we have been able to conduct security sector reform assessments in Albania, an interagency effort critical to integrating Balkan countries into the European community. We also have numerous programs targeted at countering the proliferation of weapons of mass destruction throughout the theater such as the Proliferation Security Initiative and the Global Initiative to Combat Nuclear Terrorism. But we cannot stop there. We are also supportive of efforts to pool State and Defense resources for the purpose of funding more robust, comprehensive security sector assistance programs to respond to emergent challenges and opportunities, as originally proposed by Secretary Gates. This would greatly aid our efforts to ensure interoperable, deployable NATO forces. Realizing the vision of the Lisbon Summit, a NATO with robust interoperable Article V and expeditionary capabilities requires U.S. support with training and equipment for newer NATO allies and partners. With greater flexibility, these authorities can achieve greater strategic goals in support of our theater and national objectives.

European Command and Allied Command Operations continue to serve as a transatlantic bridge that unites the United States and our partners in Europe. We are building and strengthening relations with our European partners that will help ensure the security of the United States at home and abroad. As President Barack Obama said at the recent NATO

Summit, "Our relationship with our European allies and partners is the cornerstone of our engagement with the world, and a catalyst for global cooperation." Indeed, we are truly

"STRONGER TOGETHER"

End Notes

1 The Brookings Institution, The Afganistan Index, http://www.brookings.edu/foreign-policy / afghanistan-index.aspx, 31 August 2010, 36.

2 The International Council on Security and Development (ICoS), Afghanistan: The Relationship Gap, July 2010, 14.

In: Combatant Commands in the United States Armed Forces
Editors: Quincy N. Butler and Harris A. Simmons
ISBN: 978-1-61942-924-6

Chapter 5

STATEMENT OF ADMIRAL JAMES A. WINNEFELD, JR., U.S. NAVY COMMANDER UNITED STATES NORTHERN COMMAND AND NORTH AMERICAN AEROSPACE DEFENSE COMMAND BEFORE THE SENATE ARMED SERVICES COMMITTEE*

5 April 2011

Chairman Levin, Senator McCain, distinguished members of the committee, thank you for the opportunity to share my thoughts on the missions and focus areas of U.S. Northern Command (USNORTHCOM) and North American Aerospace Defense Command (NORAD). Let me begin by saying how impressed I am by the talented cadre of professionals—from the active and reserve components of all Services of the U.S. military, DOD civilians, our Canadian partners, and representatives from 68 different civilian organizations—who work in and about my headquarters in Colorado Springs each and every day. It is truly humbling to work with such a great team.

I'm also very encouraged by the strong partnership we have with the National Guard; they are essential to operational success across the full spectrum of our missions. We have over 40 full-time National Guard positions in USNORTHCOM, including my Deputy Commander, Lieutenant General Frank Grass, from the Missouri Army National Guard, and on any given day, the number of Guardsmen in our headquarters is around one hundred. The Army National Guard provides the bulk of personnel for air defense capabilities protecting our Nation's capital. In addition, they provide all of the manning at our Ground-Based Interceptor sites in support of missile defense. They are also currently developing additional capabilities to take on a much larger role in support of consequence management in the aftermath of a chemical, biological, radiological or nuclear attack. And finally, the Air National Guard provides the majority of NORAD's operational force for Air Sovereignty

* These remarks were delivered as testimony given on 5 April, 2011. Admiral James A. Winnefeld, Jr., U.S. Navy Commander, United States Northern Command and North American Aerospace Defense Command before the Senate Armed Services Committee.

Alert missions. Simply said, we could not do our missions without the National Guard, and I'm very pleased with the positive trajectory of our relationship.

We in North America face a changing world that presents us with many challenges. These include violent extremists, proliferation of weapons of mass destruction, rogue nations, traditional competitor states, transnational criminal organizations, insecurity within the global commons, economic distress, natural disasters, emerging infectious diseases, and the effects of climate change. Each of these challenges poses a potential threat to the United States, Canada, and our regional partners and is pertinent to the missions of USNORTHCOM and NORAD. Fortunately, we also enjoy great strengths and are presented with opportunities born of our nations' ideals, ideas, and resources and those of our partners. Today I will describe how we plan to capitalize on these opportunities to overcome the challenges we face.

The mission statements for USNORTHCOM and NORAD reflect the language in the *Unified Command Plan* and the *NORAD Agreement.*

USNORTHCOM Mission: United States Northern Command conducts homeland defense, civil support, and security cooperation to defend and secure the United States and its interests.

NORAD Mission: North American Aerospace Defense Command conducts aerospace warning, aerospace control, and maritime warning in the defense of North America.

Accomplishing these missions demands a diverse array of disciplines and activities within my headquarters. Accordingly, and in order to assist me in allocating my two commands' time and resources, I have grouped our activities into the following eight focus areas:

Counterterrorism and Force Protection

Because violent extremists present a threat that currently exhibits both the capability and the intent to attack our nation, Counterterrorism and Force Protection continue to be a vital focus area for USNORTHCOM and NORAD.

Within the confines of our borders, current laws, policies, and democratic traditions and practices properly restrict most counterterrorism activities to civil authorities. As a result, USNORTHCOM's principal role is to synchronize joint force protection and ensure military infrastructure across our area of responsibility is properly postured to mitigate and prevent potential terrorist attacks. However, we are fully aligned within the federal government's counterterrorism network and play a supporting role—assisting with information sharing and remaining prepared to supply military-unique capabilities or to enhance civilian capacity when directed by the President or Secretary of Defense.

To help prevent acts of terrorism, we are working to improve information sharing to better position ourselves to preemptively detect and protect against these threats, particularly in regard to our military bases and other infrastructure. We are fully implementing the relevant recommendations of the Department of Defense Independent Review Related to Fort Hood, and have made progress over the last year in our ability to rapidly disseminate threat information to DOD installations when required. In the wake of a terrorist event, we are prepared to support civil authorities, as directed, to assist in mitigating the consequences.

COUNTERING TRANSNATIONAL CRIMINAL ORGANIZATIONS

The United States and Mexico— and many of our other Western Hemisphere partners—are confronting serious security and public health challenges driven by transnational criminal organizations (TCOs) responsible for illicit trafficking of drugs, human beings, money, and weapons.

The facts are daunting. Over 34,000 Mexicans have lost their lives in the last four years, including a dozen mayors in 2010. Murder, kidnapping, extortion, and other crimes have intimidated large segments of the Mexican populace, primarily, but not exclusively, along our shared border. In some areas, the TCOs have muzzled the media and chased away businesses.

Meanwhile, on the U.S. side of the border, the Drug Enforcement Administration estimates that Mexican TCOs operate in over 230 U.S. cities. Each year illicit drug-related deaths number in the thousands, and treatment center admissions and emergency treatment facility visits both exceed a million. The annual direct cost for treatment, prevention, interdiction, and local law enforcement of drug abuse exceeds $52 billion. These and other consequences of drug abuse, including lost productivity, the impact on the criminal justice system, and the environmental impact resulting from the production of illicit drugs are estimated to cost our nation nearly $181 billion annually.

The TCOs are vicious, well-financed and heavily armed, due in no small part to cash and weapons smuggled across our southern border. They have diversified their businesses and are increasingly sophisticated in their methods.

By fighting one another and the government for the impunity to pursue their illicit trade, the TCOs are confronting Mexico with a complex, but not unprecedented, blend of trafficking activities and challenging security problems. I am profoundly impressed by the determination and courage of the Mexican government, the various Mexican security forces, and the Mexican people in taking on this challenge. They know this is about the long-term future of their country.

The complex challenges associated with defeating the TCOs and the abundant opportunities for progress all underscore the vital importance of our close relationship with Mexico. In my ten months as the Commander of USNORTHCOM, I have observed the Mexican security forces work with increasing effectiveness against the TCOs, gradually achieving success in unraveling these organized crime networks. The Mexican army and Navy have been drawn into this struggle due to the severe threat it poses to Mexico's security and prosperity, and are working hard to overcome several important challenges. First, they must operate under the legal restrictions to which any democratic nation's military must adhere when operating within its own territory. Second, they are temporarily transforming from a traditional force to one that can be called upon to confront threats from domestic criminal enterprises. Third, they are working hard to build interagency teamwork, with budding indications of success. Finally, and despite a very information-savvy adversary, they are determined to respect human rights during their operations, and have been eager to gain insights from our own hard-earned lessons in this area. None of these challenges are foreign to our own military. And the fact that many of the families of these security forces reside in the same areas where TCO violence is greatest only heightens the respect we have for our Mexican partners.

In keeping with U.S. government policy and the law, we are working closely with the country team in Mexico City and other key stakeholders to assist the Mexican government in defeating the TCOs. Building on the momentum established by our civilian leadership, the Mexican and U.S. militaries have forged a cooperative relationship based upon mutual respect, professionalism, and reciprocity.

With all our partners, we have stressed the many dimensions of the solution—including building strong and resilient communities, forming more robust judicial institutions, establishing a 21st-century border, and directly disrupting the TCOs themselves. Regarding the latter, we are working with our Mexican military partners to enhance their materiel capability and capacity, as well as sharing our own operational insights. In so doing, we carefully emphasize the sovereignty, dignity, and capability of a proud Mexican nation, which recently celebrated 200 years of independence.

Another important partner in this effort is the Western Hemisphere Institute for Security Cooperation (WHINSEC), which provides an effective mechanism to build relationships with militaries throughout the hemisphere and to influence a positive trajectory on human rights. In June and December, I participated in the WHINSEC Board of Visitors meetings, and I have been impressed by the quality of WHINSEC's faculty and students and the foundation of respect for human rights upon which its various curricula rest. In FY10, there were 108 Mexican students at WHINSEC and I would like to see that number grow. Moreover, it is an important message both to and from Mexico that WHINSEC's Assistant Commandant is a Mexican officer.

USNORTHCOM also provides a considerable amount of support to our interagency partners operating on the U.S. side of the border. Such support includes construction of personnel barriers, roads, and bridges; air and ground transportation; intelligence support; and training in and fusion of intelligence and operations. We have begun to combine the intelligence assets at Joint Task Force North in El Paso, Texas with the interagency El Paso Intelligence Center to enable greater unity of effort. We are working closely with Customs and Border Protection (CBP) to share some of our lessons learned from ten years of countering a different irregular threat overseas, including a close partnership in CBP's Tucson Sector. Additionally, over the past year we employed multiple sensors, including radar, forward-looking infrared, as well as manned and unmanned aerial surveillance in support of the U.S. Border Patrol's counternarcotics operations on both the northern and southern borders.

In support of our northern border, last year Joint Task Force North dedicated 22% of its available resources to supporting law enforcement agencies in securing the northern border. Our support in this region included eight operational missions that provided aerial reconnaissance, ground-based radar, and ground-based sensor support; eight mobile training teams that taught targeting and intelligence courses; and one intelligence analyst who provided intelligence expertise.

From a southern border perspective, recently the aerial reconnaissance support we provided under existing counterdrug authorities assisted in the apprehension of the suspected killers of U.S. Border Patrol Agent Brian Terry. From November 2010 to January 2011, this platform assisted in the interdiction of 17,000 pounds of marijuana.

DEFENSE SUPPORT OF CIVIL AUTHORITIES

The complexity of our nation's response to natural and man-made disasters presents ample challenges and opportunities for improvement. We can respond relatively quickly to events with available DOD capability—when called upon by civil authorities. We are looking for ways to eliminate barriers to speed, enhance cooperation with our mission partners, and lean forward with Governors and the National Guard to advance understanding and teamwork.

With our National Guard partners, we are successfully exploring new ways to close an historical gap in philosophy regarding command and control of federal forces operating in support of a state in the wake of a disaster. Over the past year, together we made significant progress on an initiative, supported by the Secretary of Defense and the Council of Governors, to prepare "dual status commanders" to achieve true unity of effort between state and federal military forces in response to a natural or man-made Stafford Act incident or Economy Act event. This initiative is transforming the way we do business together. I look forward to the next National Level Exercise this spring, in which we will have the opportunity to thoroughly examine the progress we have made over the last year in unity of effort using a New Madrid Seismic Zone earthquake scenario.

As a signal of our deepening relationship, USNORTHCOM for the first time hosted Adjutants General from 54 U.S. states and territories attending the National Guard Bureau Senior Leader Conference in October 2010. The conference focused on improving understanding, fostering relationships, and furthering collaboration between USNORTHCOM and the National Guard.

In addition, I have had the honor of participating in two Council of Governors meetings. The bipartisan Council of Governors was directed by Congress in the National Defense Authorization Act for Fiscal Year 2008 and established by President Obama by Executive Order on 11 January 2010 to advise the Secretary of Defense, Secretary of Homeland Security, and the White House Homeland Security Council on matters related to the National Guard and civil support missions. These meetings have proved to be a key forum for progress in ensuring unity of effort in responding quickly in the event of disasters or other emergencies that affect the American people.

As a Combatant Commander, I am a strong advocate for the Reserve Component. I firmly believe our nation needs a strong and well-equipped Guard and Reserve force. As such, I urge Congress to fully fund the Fiscal Year 2012 President's Budget request for Reserve and National Guard capabilities for both federal and non-federal roles.

We continue to tighten our already-close relationship with the Department of Homeland Security and the Federal Emergency Management Agency (FEMA), with frequent senior leader exchanges and planning efforts. I also recently directed establishment within my staff of a cadre of Regional Desk Officers charged with working with the Defense Coordinating Officers associated with each FEMA Region, as well as a host of other partners, to bring better understanding and coordination of disaster planning between my headquarters and the various state and federal agencies.

Finally, we are closely examining the role USNORTHCOM would play in response to a cyber attack in order to synchronize our efforts with U.S. Strategic Command and U.S. Cyber

Command. We view our role as assisting the lead federal agency in mitigating the physical effects of such an event, while staying close to our partners working in the cyber domain.

Chemical, Biological, Radiological, or Nuclear (CBRN) Consequence Management

Managing the aftermath of a CBRN event would be exceptionally challenging due to the potential scope of an event, the specialized skills required, and the general lack of knowledge among our population of the relative hazards associated with such an incident. USNORTHCOM has a key leadership role in ensuring our nation is prepared to succeed in this mission area. The Quadrennial Defense Review directed a transition to a CBRN Consequence Management Enterprise with more responsibility resident within the National Guard, more lifesaving capability, and faster response times. I am convinced this is the right path for this capability, which will be in high demand if such an event ever occurs. USNORTHCOM and our ground component command, U.S. Army North, are working closely with the National Guard Bureau and the Joint Staff to assist in making the new enterprise operational. The first two state Homeland Response Forces and the federal Defense CBRN Response Force, or DCRF, will stand up this fiscal year. While we collectively have much work to do to bring this project to fruition, I am pleased to report that together with our partners we have leveraged excellent teamwork across the board in this effort, and are committed to ensuring a smooth transition from existing to planned capability in this area.

Maritime Warning and Control

We remain concerned with potential threats in the maritime environment—whether presented by nation-states, extremists, or a natural event. Moreover, opportunities abound for shared awareness and control and much remains to be done to both clarify and energize NORAD's mission area of Maritime Warning.

One of our key projects this year is to strengthen our day-to-day maritime componency relationships for USNORTHCOM and NORAD to enable improved planning, maritime domain awareness, training, theater security cooperation, and execution of homeland defense and security operations—and to do so without requiring additional investment.

Aerospace Warning and Control

This continually evolving mission is NORAD's central focus, and it maintains the same importance to the United States and Canada that it has for over 52 years. It's a real privilege to have 122 Canadians in my headquarters as part of a team that reflects the overall vitality of the relationship between our two nations.

We continue to challenge our assumptions in this arena to ensure we are accounting for potential changes in threats before they occur, while offering our two nations our best thinking on how to execute this mission as efficiently as possible.

We view this problem as a spectrum of potential threats with varying capabilities and intentions. Despite recent improvements in U.S.-Russian relations that reflect a dramatically reduced likelihood of conflict, we maintain our vigilance regarding the high-end threat to the United States and Canada because of enduring and continually improving Russian capability. However, our principal concern remains potential extremist intent to again use civil aircraft, ranging from commercial airliners to general aviation aircraft to ultra-lights, as a means for employing terrorist tactics. While we have made tremendous progress in our efforts to ensure a 9/11-type of event can never occur again, the threat continues to search for ways to exploit potential vulnerabilities. Accordingly, we search every day for new ways to improve our tactics, techniques, procedures, and technical capability to enable us to execute the detect-assess-engage sequence as effectively as possible.

To do so, we need the right capability in the right place at the right time. We are analyzing our future Air Sovereignty Alert requirements and sharing this analysis with the leadership of the U.S. Air Force and the Air National Guard. We are focusing our analytical effort on exactly what our Air Sovereignty Alert force is intended to protect, against what types of threats, under what conditions, and at what cost. As part of this effort, we are identifying capability gaps to the Joint Staff and the Services that we believe are required to defend against an evolving threat and contributing to an examination of means by which some of these gaps, which remain classified, might be filled. In addition, we are preparing a Report to Congress on the Air Sovereignty Alert mission as directed by Section 333 of the National Defense Authorization Act of Fiscal Year 2011.

We also need to ensure we have the right capacity to execute the Air Sovereignty Alert mission—not too much and not too little. Given the recently announced slip in the F-35 program, NORAD is working closely with and counting on the Air Force to ensure we have adequate resources to sustain our mission. This is critical because seven of our Air Sovereignty Alert sites fly older model F-16s, which are currently scheduled to reach the end of their service life between 2020 and 2023.

Another area we are watching closely is the pace of wind farm development. Increases in the number of wind farms raise the likelihood that radar signals vital to our ability to protect the national airspace will be obstructed. We believe enabling the construction of alternative energy sources and conducting our national air defense mission are not mutually exclusive as long as we exercise due diligence in assessing the impact of potential projects. To this end, we have developed a more mature process for evaluating the impact of wind farms on national security. Multi-departmental cooperation is required to develop the policy, technical solutions, and future surveillance infrastructure that will provide both national security and renewable energy at the same time.

We are also fostering a more collaborative relationship with the Russian Federation. In August 2010, NORAD and the Russian Federation completed an historic first—a cooperative three-day, live-fly exercise designed to establish clear communication processes that would allow our two forces to work together during a real crisis. The exercise, VIGILANT EAGLE, was an international air terrorism scenario exercised over the Pacific Ocean consisting of forces from the United States and Russia responding to the simulated hijacking of a B-757 en route to and from the Far East. The exercise scenario created a situation that required both the Russian Air Force and NORAD to launch or divert fighter aircraft to investigate and follow a hijacked airliner, with a focus on shadowing and coordinated hand-off of monitored aircraft between fighters of our two nations. This exercise was an overwhelming success and has helped create an environment for further cooperative efforts.

MISSILE DEFENSE

Several nations are developing the capacity to target North America with ballistic missiles capable of carrying weapons of mass destruction in the belief those weapons will give them more freedom of action. Thus far, the United States is pacing the threat, but a lack of certainty of threat intentions and capabilities demands vigilance and agility. We focus on three imperatives in order to perform our missile defense mission:

- Execute the ballistic missile defense mission with precision using the existing Ground-Based Interceptor (GBI) fleet located at Vandenberg Air Force Base, California and Fort Greely, Alaska
- Develop realistic training simulations and constantly train as we intend to fight
- Assist the Missile Defense Agency (MDA) and U.S. Strategic Command as they continue concurrent research and development activities to improve capability

I have gained increased confidence in the existing ballistic missile defense system's ability—including our sensors, weapons systems, and highly trained operators—to defend against current limited threats. Nonetheless, I would like to see a more robust and redundant architecture for sensor and command and control nodes. It is critical that we continue to ensure our sensor network provides adequate warning and targeting information, that we test the entire system to verify its reliability and validate ongoing improvements, and that we remain vigilant to ensure our capabilities remain ahead of the evolving threat. Accordingly, I fully support the Department's recent decision to keep the GBI production line open until at least 2016, as well as consideration to procure additional GBIs in light of recent flight test results.

As we continue to monitor other nations' advancements in their long-range missile programs, I am optimistic the Administration's Phased Adaptive Approach to Ballistic Missile Defense will add another layer of defense for the homeland if future conceptual programs envisioned to support this approach materialize and mature. In the meantime, MDA has begun to demonstrate a 2-stage GBI capability that I believe could contribute to U.S. homeland defense if a more mature threat appears more rapidly than we had originally projected.

I remain alert to our ability to defend the nation against the potential future threat of cruise missiles and other less conventional forms of air attack from off our coasts. This is a complex problem space that is defined by: the spectrum of capabilities and intentions of potential threats; our ability to generate indications that these capabilities or intentions might be changing; and the costs, timelines, and capabilities of varying levels of potential defensive postures. Presently, we are aware of no threat possessing both capability and intent to conduct such an attack. However, we must remain vigilant and prepared to adjust our posture should a threat combination emerge that changes this equation. Clearly, a strong deterrence posture is one of our most important hedges against such a threat. USNORTHCOM and NORAD are drafting a Report to Congress on this issue as directed by the House Report to Accompany the National Defense Authorization Act for Fiscal Year 2011 (House Report 111-491).

THE ARCTIC

The geopolitical importance of the Arctic has never been greater, because as far as we know, the natural environment in the Arctic in civilized times has never changed faster.

Up to 25% of the world's remaining undiscovered oil and natural gas deposits may lie beneath the Arctic ice cap. While most experts believe it will be some time before commercial Arctic shipping routes through the Northwest Passage and the Northern Sea Route see a significant increase in volume, some countries and commercial interests are actively testing the waters and making plans to increase their activity. We have seen a marked increase in Arctic ecotourism, and its attendant safety concerns, including the grounding of a cruise ship in the Northwest Passage last summer.

Because these changes involve a complex mosaic of issues, challenges, and opportunities, and because a peaceful Arctic is central to the continued safety and security of the United States, I have elevated the Arctic to the status of a key focus area. We are crafting a Commander's Estimate on the Arctic for use within DOD, and my commands are examining how we can best support our interagency partners in this region with search and rescue assets, humanitarian assistance, disaster response capabilities, and support to law enforcement. We are also working hand-in-hand with Canada Command as a vital partner to produce a concept of operations regarding how we would partner in the Arctic to ensure our efforts are coordinated and that we pursue complementary rather than redundant capabilities in accordance with our respective national direction.

Regarding capabilities, we are maturing our understanding of our gaps in this unique environment. We face shortcomings in all-domain awareness, communications, infrastructure (to include a deepwater port), mobility (to include adequate national icebreaking capability), search and rescue enabling capabilities, Arctic Ocean charting, and the ability to observe and forecast Arctic environmental change.

The good news is that cooperation is on the rise in the Arctic, and we must continue that trajectory using the array of mechanisms available to us, such as the Arctic Council, the International Maritime Organization, and the United Nations Convention on the Law of the Sea. I would like to add my voice to those of the Secretary of Defense, Chairman of the Joint Chiefs of Staff, and Chief of Naval Operations in urging the Senate to ratify the latter. Becoming party to the Convention would protect and advance U.S. interests in the Arctic by bolstering our national security, securing U.S. rights over extensive marine areas, and giving the United States a seat at the table when our vital interests are at stake—without abdicating any sovereignty.

CONCLUSION

It is a privilege to be the military commander charged with the diverse array of missions to protect our vital interests in our homeland, whether deterring or defeating a direct attack, or supporting civil authorities in disrupting and defeating TCOs, or supporting other civil authorities in responding to disasters.

You have a great team in USNORTHCOM and NORAD—Americans and Canadians serving side-by-side. We are proud to serve together and as we do, we remember the vital importance to both the United States and Canada of the NORAD partnership, as well as the broad spectrum of missions demanded of USNORTHCOM. We pledge to you—the U.S. Congress and the American and Canadian people—to give each one our best effort. Thank you.

In: Combatant Commands in the United States Armed Forces ISBN: 978-1-61942-924-6
Editors: Quincy N. Butler and Harris A. Simmons

Chapter 6

STATEMENT OF ADMIRAL ROBERT F. WILLARD, U.S. NAVY COMMANDER, U.S. PACIFIC COMMAND BEFORE THE SENATE ARMED SERVICES COMMITTEE ON U.S. PACIFIC COMMAND POSTURE*

12 April 2011

INTRODUCTION

Chairman Levin, Senator McCain and distinguished members of the committee, thank you for this opportunity to provide an update on United States Pacific Command (USPACOM). Today is my second testimony as the Commander of USPACOM. Since the 2010 hearings, much has happened in this increasingly important region. I have travelled throughout the Asia-Pacific visiting a dozen different countries—some multiple times—and gained valuable insight into the extensive contributions by our regional allies and partners.

What follows is my assessment of the strategic environment and the role of USPACOM forces assigned throughout the region. Included in this testimony is a discussion of the military construction (MILCON) programs that enable these forces to perform their mission effectively while sustaining the quality of life our servicemen and women deserve.

STRATEGIC ENVIRONMENT IN THE ASIA-PACIFIC

The USPACOM Area of Responsibility (AOR) is vital to U.S. national interests. It spans half the earth and is home to more than three billion people living in three dozen nations—five of which are allied with the U.S. and many more of which are important economic and

* These remarks were delivered as testimony given on April 12, 2011. Admiral Robert F. Willard, U.S. Navy Commander, U.S. Pacific Command before the Senate Armed Services Committee on U.S. Pacific Command Posture.

security partners. The region contains the world's three largest economies, and almost one-third (over $1 trillion) of U.S. two-way trade in goods and services is with nations in the region. Moreover, much of the world's trade and energy that fuels the world economy moves on Asia's sea and air lines of communication.

The vastness of the region makes permanent and rotational U.S. force presence essential to enabling security and strategic deterrence throughout the region while protecting and defending the homeland. Military construction provides necessary facilities for new weapon systems, supports the Services' evolution to become more efficient and effective, offers warfighters and their families quality-of-life facilities while at home, and renovates existing facilities that are beyond their useful lives. Thus the MILCON projects in this testimony enhance the capabilities of USPACOM forces that underpin security in this increasingly important and dynamic region.

While the region remains relatively secure and stable, the strategic environment also includes traditional and asymmetric challenges that drive the need for forward presence and the subsequent MILCON recommendations in this testimony. Sustaining the conditions that have underpinned unprecedented prosperity for over six decades remains challenging for a variety of reasons, including the following:

- The threat to the United States and its allies posed by North Korea's nuclear and missile capabilities, its proliferation of weapons of mass destruction and associated technologies, and its potential for instability
- Transnational violent extremist organizations (VEOs) undermine stability and threaten traditional allies and emerging partners
- China's significant military modernization associated with its unclear intent
- Territorial disputes, and increasingly assertive actions to resolve them, present the potential for conflict and instability
- Increasingly persistent and sophisticated cyber threats that challenge unencumbered operations.
- Transnational criminal activity—to include piracy and trafficking in narcotics and persons— that rejects the rule of law and challenges international order
- Humanitarian crises such as pandemics and famines, as well as natural disasters such as tsunamis, earthquakes, and volcanoes
- Environmental degradation caused by poor resource management, the pillaging of natural resources, and disputes over resource sovereignty

Despite these many challenges, the region remains one of immense opportunity for peaceful growth, cooperation, and prosperity. Realizing such opportunity relies upon continued U.S. ability and willingness to underwrite security, extend deterrence, and protect the global commons upon which the region's livelihood depends. U.S. military strength, presence, and engagement provide the means to ensure security and peace and avoid confrontation and conflict. Secretary of Defense Gates emphasized this point in Singapore in June 2010: ―The strength of the U.S. commitment and deterrent power will be expressed through the continued forward presence of substantial U.S. forces in the region.[1]

USPACOM thus embraces a theater strategy that leverages an evolving force posture. In concert with other government agencies, this posture is designed to simultaneously hedge against traditional and asymmetric challenges as well as advance alliance and partner-nation relationships. Extensive analyses clearly indicate a need to build an integrated posture framework that prioritizes adjustments by maximizing strength, balancing and biasing disposition, and sustaining readiness in all sub-regions (Northeast Asia, Southeast Asia, South Asia, and Oceania).

Regional Force Posture Assessment

Broadly, the U.S. military in the Asia-Pacific enjoys freedom of action, numerous dependable allies and partners, and ready access to theater forces (e.g., Hawaii, Guam, and Japan) as well as to global and continental (especially West coast and Alaska garrisoned) forces. Other existing posture strengths include the collocation of command elements in Hawaii, pre-positioned stocks, maintenance support, and several distributed, forward sub-commands.

Current force posture throughout the Asia-Pacific remains heavily influenced by post-World War II- and Cold War-era basing and infrastructure. While maintaining access and capabilities in Northeast Asia remains essential, attaining better access to and support from Allied and partner nations in South and Southeast Asia is increasingly important. As Chairman of the Joint Chiefs of Staff Admiral Mullen stated in the 2011 National Military Strategy, ―our presence and alliance commitments remain the key to preserving stability in Northeast Asia, [and] we must also invest new attention and resources in Southeast and South Asia.[1]

USPACOM strives to continuously optimize its force posture to meet emerging 21st-century conventional and asymmetric threats. Nevertheless, forward forces remain potentially vulnerable to cyber attack on networks and logistics systems. Also, growing anti-access and area-denial challenges make joint capabilities and cooperation imperative; further development of the Air-Sea Battle concept will establish a better institutionalized method to address this threat.

NORTHEAST ASIA

Northeast Asia is a complex region with distinct challenges and a wealth of opportunities to positively influence the security environment. It contains many of the most significant economies and militaries in the Asia-Pacific and the world, including Japan, the Republic of Korea (ROK), China, and Russia. Strong U.S. presence, basing, access, and support in this important sub-region serves to not only protect the homeland but also furthers U.S. interests regarding allies, partners, extended deterrence, and treaty obligations.

Northeast Asia is home to North Korea, one of the most likely and persistent threats to the U.S. and its allies. This nation presents a multifaceted problem set for the U.S.-ROK Alliance, the region, and the international community. In addition to the conventional threat it poses to the ROK, its nuclear program, missile development, proliferation activities, provoke-bargain-cheat cycles, add to North Korea's capacity to disturb peace and stability throughout the region and globally.

Japan and the ROK are strong U.S. allies that host U.S. forces, enjoy U.S. extended deterrence, and stand with the U.S. in containing North Korea's aggression and meeting other regional and global security challenges.

Japan

Rapid response in the Asia-Pacific region hinges on flexibility and forward basing of military forces. Despite some recent deliberation related to U.S. basing realignment in Japan, the military relationship and the overall Alliance remain strong. As the Alliance enters its 51st year, it endures as relevant and as a key component of security in Northeast Asia and the larger Asia-Pacific. The U.S. and Japan continue to share similar security interests; therefore continued collaboration is vital to address challenges that include DPRK, humanitarian assistance and disaster relief (HA/DR), and support for freedom of action in the maritime domain as well as in space and cyberspace.

Defense Policy Review Initiative (DPRI). To transform and realign U.S. forces in Japan, the government of Japan (GOJ) and the U.S. government agreed in 2005 to implement changes that would allow USPACOM forces to adapt to the changing regional and global security environment. The Defense Policy Review Initiative (DPRI) remains a key transformation goal of the U.S.-Japan Alliance and USPACOM remains committed to its implementation. Major realignment elements include relocating two U.S. air bases from urbanized to rural areas, co-locating U.S. and Japanese command and control capabilities, deploying U.S. missile defense capabilities to Japan in conjunction with their own deployments, and improving operational coordination between U.S. and Japanese forces.

DPRI implementation, in large measure, is on track due to the significant contributions provided by the GOJ. The issue lacking progress and most familiar to members of Congress is the development of a Futenma Replacement Facility and the subsequent move of approximately 8,000 Marines to Guam. While this lack of progress is disappointing, it does not undermine the ability of USPACOM to maintain trained and ready forces in the region or to fulfill U.S. treaty obligations to Japan. Meanwhile, U.S. forces will continue to operate from the existing facility at Futenma Marine Corps Air Station while USPACOM works with Department of Defense leadership and the Japanese to find a solution acceptable to both nations.

Guam

Like the Okinawa-based Marines, Guam-based forces offer the response and deterrence required to enhance security to meet U.S. interests and fulfill commitments to regional U.S. allies. Per the original agreement, U.S. MILCON funds will also contribute to the facility and infrastructure construction costs on Guam to support the relocating Marines and their dependents. In addition, investments will be needed to improve off-base Guam infrastructure, such as roads and bridges critical to the successful movement of materiel to the planned Marine Corps installation.

The FY2011 National Defense Authorization Act did not include $320 million for three projects that were submitted in the President's Budget that support the move of Marines from

Okinawa to Guam. The reasons for not funding the three priorities are explained in the Joint Explanatory Statement of the Committee on the Armed Services. Currently, the U.S. and Japanese governments continue to reaffirm agreements and provide associated funding, including the second Japanese funding increment of $498 million to support the move of Marines to Guam.

Also, recent approval of the Programmatic Agreement in the National Historic Preservation Act Section 106 process will enable DPRI-related construction to begin.

Therefore, for FY2012 the Marines are resubmitting requests for North Ramp Utilities improvement and Finegayan Water Utilities & Site Preparation for a total of $156 million to support U.S. contribution to the 2006 Realignment Roadmap. The Finegayan project was reduced by $70 million to include only the water distribution infrastructure. The funding for these projects is critical to the success of the DPRI roadmap and an important signal of U.S. commitment to treaty obligations with Japan.

The Republic of Korea

The U.S.-ROK Alliance remains strong and is critical to U.S. strategy and military force posture in Northeast Asia and the region. This Alliance is also transforming to ensure it matures as an effective contributor to Peninsula and regional security.

Current initiatives are underpinned by the transition of military wartime operational control of the ROK to its own forces during peacetime and war by the end of 2015. Overall, this transformation effort is designed to improve responsiveness to an evolving peninsular, regional, and global security environment. To this end, the transformation must ensure U.S. forces, which are stationed on the Korean Peninsula to deter North Korea's aggression and respond to North Korea's hostility, are also accessible for regional and global employment. Further, the transformation will ultimately support the ROK military's continued development of extra-peninsular capabilities and capacities, and enable additional contributions like those seen recently in the Gulf of Aden combating piracy, providing peacekeeping forces to Haiti and Lebanon, and in Afghanistan leading a Provincial Reconstruction Team.

USPACOM posture transformation in the ROK entails multiple efforts. Projects for the Land Partnership Plan and the Yongsan Relocation Plan include relocating U.S. troops out of downtown Seoul, returning the majority of Yongsan Army Garrison to the ROK government, and consolidating U.S. forces into two primary hubs south of Seoul. This effort sustains a strong U.S. combat presence on the Peninsula while reducing force presence in major urban centers like Seoul. Of note, this realignment is funded largely by the ROK, but MILCON funds are needed for some facility construction and infrastructure developments. While the Services are not submitting MILCON projects related to transformation in FY2012, additional, important projects will be submitted in future fiscal years.

Trilateral Cooperation

Trilateral security cooperation between the U.S., ROK, and Japan also continues to improve. Building on longstanding efforts to enhance policy coordination and strategic dialogue among the three countries, Secretary Clinton met with Foreign Ministers Maehara and Kim here in Washington on December 6, 2010, to discuss ways to address enduring and emerging challenges. When the devastating earthquake and tsunami hit Japan, the ROK demonstrated unprecedented support by immediately dispatching rescue teams and sending emergency supplies. That said, there are outstanding political, military, and historical impediments to the Japan-ROK relationship that must be overcome. The shared values and common interests, as well as the financial resources, logistical capabilities, and planning potential to address complex contingencies throughout the region, make expanding this trilateral partnership a worthy effort. Encouraging steps are also apparent in the burgeoning trilateral relationships between the U.S., Japan, and Australia, and between the U.S., Japan, and India.

The People's Republic of China (PRC)

China's rise will largely define the Asia-Pacific environment in the 21st century. As noted in the 2010 National Security Strategy, ―We welcome a China that takes on a responsible leadership role in working with the United States and the international community to advance priorities like economic recovery, confronting climate change, and nonproliferation. We will monitor China's military modernization program and prepare accordingly to ensure that U.S. interests and allies, regionally and globally, are not negatively affected. More broadly, we will encourage China to make choices that contribute to peace, security, and prosperity as its influence rises.

In support of this broader national strategy toward China, forward-postured USPACOM forces are focused on deterrence and reassurance missions as they apply to China and U.S. allies and security partners in the region. In addition, USPACOM's interactions with China assist the Administration's broader goals by contributing to an overall military-to-military relationship that is healthy, stable, reliable and continuous. Such a relationship is important to avoid misperception, miscommunication, and miscalculation while it expands opportunities for cooperation where our security interests overlap. However, our military relationship with China continues to suffer from an on-again/off-again cycle of interactions which limits its ability to accomplish the above tasks. China suspended bilateral military relations following our arms sales to Taiwan in January 2010 and restarted them in fall 2010. We look forward to continuing the progress made in recent months which includes Secretary Gates' successful visit in January of this year.

China's Military Modernization Program

Beginning in the mid-1990s, China's peacetime military modernization program has progressed at a rapid rate. While force modernization is understandable in light of China's growing regional and global roles and accompanying requirements, the scope and pace of its modernization without clarity on China's ultimate goals remains troubling. For example, China continues to accelerate its offensive air and missile developments without corresponding public clarification about how these forces will be utilized. Of particular concern is the expanding inventory of ballistic and cruise missiles (which include anti-ship capability) and the development of modern, fourth- and fifth-generation stealthy combat aircraft. In conjunction, China is pursuing counter-space and -cyber capabilities that can be used to not only disrupt U.S. military operations, but also to threaten the space- and cyber-based information infrastructure that enables international communications and commerce.

Absent clarification from China, its military modernization efforts hold significant implications for regional stability. The region is developing its own conclusions about why the People's Liberation Army (PLA) continues to expand its ability to project power outside China's borders, and to range both U.S. forces and U.S. allies and partners in the region with new anti-access and area-denial weaponry. Of growing concern is China's maritime behavior. China's recent official statements and actions in what Beijing calls its ¯near seas" represent a direct challenge to accepted interpretations of international law and established international norms.

While China does not make legal claims to this entire body of water, it does seek to restrict or exclude foreign, in particular, U.S., military maritime and air activities in the ¯near seas" - an area that roughly corresponds to the maritime area from the Chinese mainland out to the ¯first island chain" (described, generally, as a line through Japan, Taiwan, Philippines, and Indonesia) and including the Bohai Gulf, Yellow Sea, East China Sea, and South China Sea. Chinese naval and maritime law enforcement vessels have been assertive in recent years in trying to advance China's territorial claims in the South China and East China Seas which has resulted U.S. partners and allies in East Asia seeking additional support and reassurance to balance and curb the Chinese behavior. Many of China's maritime policy statements and claims stand in contrast to the United Nations Convention on the Law of the Sea (UNCLOS). The U.S. has consistently sought the appropriate balance between the interests of countries in controlling activities off their coasts with the interests of all countries in protecting freedom of navigation. China has questioned whether a non-party may assert such rights under UNCLOS, a baseless argument but one that would be removed if the U.S. was a party to UNCLOS.

The current situation in the Taiwan Strait remains stable as tensions have declined in recent years; however, the Taiwan issue remains a challenge to long-term regional stability. China refuses to renounce the use of force to resolve the Taiwan question. As China's military modernization proceeds, the cross-Strait military balance continues to shift in the mainland's favor.

U.S. policy in support of a peaceful resolution remains consistent and clear. We are committed to our one-China policy, based on the three U.S.-China communiqués and the Taiwan Relations Act. We do not support Taiwan independence, and would oppose unilateral changes, by either side, to the status quo.

Russia

Russia's Pacific armed forces have not re-emerged from their post-Cold War decline. Moscow is nonetheless attempting to reassert influence in the region through increased naval and strategic air force operations, cyberspace activities, and arms sales to the region's emerging and rising nations. That said, USPACOM enjoys a relatively positive militaryto-military relationship with Russia, especially between each country's Pacific Fleets.

Mongolia

Mongolia endures as a small but strong partner in Northeast Asia. It continues to demonstrate support for U.S. regional and global policy objectives—especially those linked to the Global Peacekeeping Operations Initiative and security operations in Afghanistan—while managing positive relations with neighbors China and Russia.

Northeast Asia Force Posture Assessment

Strategic deterrence in the USPACOM AOR is concentrated in Northeast Asia where the great majority of U.S. force posture is based. Although the overall combined strength in this sub-region is formidable, there remain areas of concern such as ballistic missile defense and regional contingency support. While both the U.S. and its Northeast Asian allies and partners have invested in the technical capabilities required to mount effective defenses against ballistic missile attack, the broader missile defense of the region remains problematic, especially in response to a large-scale attack. Further, while Taiwan's military strength is currently sufficient to deter full-scale amphibious or air assault by China, Taiwan's ability to maintain that capability will hinge, in large measure, on whether Taiwan continuously invests in upgraded systems. Lastly, while U.S. military posture in the Asia-Pacific has traditionally focused on the sea, undersea, air, and land domains, the modern realities of warfare demand that it defend against challenges to operate freely in the space and cyberspace.

SOUTHEAST ASIA

Southeast Asia is a diverse sub-region, rich in natural resources, and strategically located at the crossroads of the Pacific and Indian Oceans. It is host to a mix of democratic and authoritarian governments, prosperity and poverty, , disparate military capabilities, and significant human capital potential. Prospects for continued development, increased security, and regional integration are promising, but the sub-region presents significant security challenges. China's increasing engagement in this sub-region—which, in many cases, is aimed at supplanting U.S. influence—as well as its expansive claims to, and growing assertiveness in, the South China Sea are two notable challenges. Southeast Asia is also home to transnational challenges and threats that include violent extremism, piracy, refugee migrations, and transnational crime such as narcotics and human trafficking. Additionally, this sub-region is particularly susceptible to natural disasters such as typhoons, earthquakes, volcanic eruptions, floods, and tsunamis.

The myriad of challenges faced by U.S. allies and partners in Southeast Asia converge with U.S. interests; therefore strengthening defense partnerships is important to regional security. To further military interoperability and build regional capacity to respond to these challenges, the establishment of foundational information, logistics, and technology exchange agreements with these allies and partners is important.

In response to the growing threat posed by VEOs, USPACOM supports improving the CT capabilities with security assistance programs, executed in partnership with the

Department of State and embassy country teams, designed to build host-nation capacity and capability. Foreign Military Financing—the program for funding the acquisition of U.S. military materiel, services, and training that support regional stability goals—has been particularly important to supporting partners engaged in combating violent extremism, especially the Philippines and Indonesia. Additionally, International Military Education and Training funding has been very valuable to establishing the network of relationships that underpin cooperative success and access to leadership in Southeast Asia.

Philippines

The U.S. and the Philippines share democratic values, a long and unique history, and a desire to contribute to regional security. Underpinned by the 1951 Mutual Defense Treaty, the U.S.-Philippines Alliance remains important to U.S. strategy in the Asia-Pacific. A significant effort in the Philippines is conducted by Joint Special Operations Task Force-Philippines (JSOTF-P). This task force operates in support of the Armed Forces of the Philippines (AFP) against the VEO threat posed mainly by the Abu Sayyaf Group and Jemaah Islamaiya in Southern Mindanao and the Sulu Archipelago Region. The task force has enjoyed significant success advising, training, exercising with, and informing the AFP, and contributed to the killing or capture of a significant portion of VEO leadership, the denial of safe havens to extremists, and the enhancement of quality of life for local populations. USPACOM engages in many other ways with the Philippines—to include the Joint Staff-sponsored Exercise BALIKATAN—to help shape the AFP into a force capable of responding to various natural disasters as well as border and self-defense (particularly important regarding maritime security and sea lines of communications).

Thailand

Thailand is an important Southeast Asian ally and engagement partner, with whom USPACOM intends to further strengthen and broaden the Alliance. The U.S. and Thailand have worked together to bolster regional stability and security through cooperation on humanitarian and disaster relief, peacekeeping training, and counter-proliferation. This past year, the United States and Thailand partnered to deploy two Thai naval vessels, with U.S. Navy personnel aboard, to join Combined Task Force-151 combating piracy off the Horn of Africa. The Royal Thai Army, working with the United States, also deployed a full battalion of peacekeepers to Darfur to assist with UN humanitarian relief operations. Our relationship has even broken new ground in the field of public health, where U.S. and Thai military medical professionals have worked to combat infectious diseases from malaria to HIV. Further, U.S. forces benefit from our mutually beneficial relationship with Thailand, which allows access and engagement in an important geo-strategic location. Further the broad access granted to U.S. forces to logistical facilities, ports, airfields and training areas is crucial to enabling regional force projection and contingency response.

The premier Exercise COBRA GOLD is a Thailand-hosted USPACOM multilateral exercise that includes seven participating nations (U.S., Thailand, Japan, South Korea, Singapore, Indonesia, and Malaysia) and observers from over 15 additional nations. COBRA

GOLD is just one of many military exercises that we co-host with the Thais, each of which bolsters unique capabilities within our forces and develops key areas of interoperability with our Thai allies and other partners.

Singapore

The U.S.-Singapore relationship has matured significantly in the past six years and contributes in many ways to enhanced regional security. In 2005, Singapore and the U.S. advanced the relationship with the signing the Strategic Framework Agreement that identifies the two nations as ―Major Security Cooperation Partners.[1] Singapore hosts transiting U.S. Navy ships and deployed personnel, works with U.S. forces in Afghanistan, and plays a leading role in the multi-national, counter-piracy, Combined Task Force in the Gulf of Aden.

Indonesia

The relationship between USPACOM and Indonesia—the world's fourth most populous nation, third largest democracy, and largest Muslim-majority country—continues to mature. Following a decade of political and economic reform, Indonesia has emerged as a vibrant democracy, increasingly confident about its leadership role in Southeast Asia and the developing world, particularly in disaster risk reduction, HA/DR, and peacekeeping operations.

Indonesia, along with Japan, recently demonstrated its leadership role in building regional capacity for humanitarian assistance by co-hosting the Association of Southeast Asian Nations (ASEAN) Regional Forum Disaster Relief Exercise.

After years of limited engagement with the Indonesian Armed Forces, an increase in transparency and pace of institutional reforms have recently provided impetus to renew, advance, and broaden the U.S.-Indonesia military-to-military relationship. In recognition of Indonesia's pledges to protect human rights and advance military accountability, in July 2010, Secretary Gates noted that after a 12-year hiatus, the U.S. intends to begin a measured and gradual program of security cooperation activities with Indonesian Army Special Forces (KOPASSUS) within the limits of U.S. law.

Malaysia, Vietnam, Cambodia, Laos, Brunei and Timor-Leste

The U.S. has extensive interests across the rest of Southeast Asia, and USPACOM seeks to continue maturing military relations with Malaysia, Vietnam, Cambodia, Brunei, and Timor-Leste. Malaysia maintains a strong leadership role in the region especially combating terrorism and enhancing maritime security. Military relations with Vietnam continue to grow in areas such as disaster management, conflict resolution, personnel recovery, medical capability, and trafficking in persons. Cambodia is emerging as a strong supporter of U.S. policy in the region, and activities with Laos, Brunei and Timor-Leste have expanded over the past year.

Association of Southeast Asian Nations (ASEAN)

Focused efforts by the President and Secretaries Clinton and Gates have set the conditions for increased security cooperation with ASEAN. Accordingly, USPACOM now actively seeks opportunities to support ASEAN initiatives that complement developing coordinated, multilateral approaches to maintaining regional security. ASEAN and USPACOM have several convergent interests that include humanitarian assistance/disaster relief, maritime security, terrorism, and peacekeeping. With an evolved scope and influence, ASEAN and its offshoots (ASEAN Regional Forum, ASEAN Defense Ministers Meeting Plus, and the East Asia Summit) offer a framework to guide regional efforts to secure the commons and underpin cooperation across governmental institutions.

Burma

Burma presents a number of challenges to regional security, including violence between Burmese military forces and various domestic ethnic armies which sparks significant, refugee flows into Thailand; a maritime border dispute with Bangladesh, narcotics and human trafficking, and the potential for rapid spread of pandemic disease. Of increasing concern is Burma's role in regional weapons proliferation, evidenced by increasing defense cooperation and procurement from the North Korea, in violation of several UNSCRs. In addition, Burma's government continues to rule without respect for democracy or basic human rights The U.S.-Burma military-to-military relationship remains suspended as a matter of U.S. policy and public law.

Engagement Programs

Two significant engagement programs USPACOM employs are PACIFIC PARTNERSHIP and PACIFIC ANGEL. They are designed to enhance regional stability by building partner-nation resiliency; focus on growing multilateral capacities and capabilities in the areas of consequence management, civic assistance, and HA/DR. In 2010, U.S. Naval Ship *MERCY*, with personnel from ten nations, successfully accomplished this mission during a five-month deployment to Vietnam, Cambodia, Timor-Leste, and Indonesia. The mission treated over 110,000 patients, conducted scores of community relations projects, and completed many engineering and infrastructure projects. In 2011 PACIFIC PARTNERSHIP will return to Oceania aboard *USS CLEVELAND*. PACIFIC ANGEL, meanwhile, utilizing C-17 aircraft, cared for tens of thousands of patients and completed several medical and civic action projects in the Philippines, Vietnam, Bangladesh, and Sri Lanka.

Southeast Asia Force Posture Assessment

U.S. force posture in Southeast Asia is achieved almost entirely with deployed U.S. forces, which is a costly and inefficient way to achieve presence and conduct the requisite security and influence activities. Singapore, which allows a modest mix of U.S. service capabilities, is the only exception in the region.

Expanding U.S. military presence in Southeast Asia is a mid- to long-term prospect. In general, the Philippines remains restrictive regarding U.S. military access, except for JSOTF-P CT support which operates at the invitation of the Philippine government. Thailand has permitted consistent access to its military facilities for transient and exercise-related operations, but any additional force presence would require thorough discussions with our Thai allies.

South Asia

The South Asian sub-region is dominated by India, an emerging U.S. strategic partner, and also includes Bangladesh, Maldives, Nepal, Sri Lanka, and Bhutan. While there are prospects for enduring peace and prosperity, this sub-region is challenged by historical animosity between India and Pakistan, contested borders between India and China, and transnational threats that include terrorism, WMD proliferation, piracy, and narcotics trafficking. The sub-region is also prone to a wide array of natural disasters and lacks the resources and capabilities to effectively respond. This is exacerbated by the absence of a strong multilateral security structure.

The encroachment of piracy emanating from Somalia into the USPACOM AOR in South Asia is a growing concern. This issue poses a particular threat to the vast sea lines of communication that span the Indian Ocean. As a result of this expanding problem, USPACOM seeks to partner with India to counter this threat.

India

In July 2010 at the Asia Society, Undersecretary of Defense for Policy Flournoy characterized the U.S.-India relationship as follows: ―This bond is grounded in common democratic values and converging interests that make India and U.S. natural partners. The U.S. and India have an overarching shared interest in promoting global stability and security.[1] Given the numerous areas where U.S. and Indian interests converge, plus many shared national values, a strong India-U.S. partnership is important to future South Asian security.

Cooperation is especially noteworthy in the areas of counterterrorism, humanitarian assistance and disaster relief, and maritime security. The recent removal of Indian defense- and space-related industries from the U.S. Entity List not only recognizes India's record of responsible stewardship of sensitive technologies but further enables bilateral cooperation in areas of mutual interests.

Nevertheless, India's historic leadership of the non-alignment movement and desire to maintain strategic autonomy somewhat constrain cooperation at a level USPACOM desires. The U.S.-India relationship remains challenged by a degree of suspicion fueled by Cold War-influenced perceptions, complicated Indian political and bureaucratic processes, and the U.S.-Pakistan relationship.

Relations with both India and Pakistan are vital to U.S. national interests and cannot be addressed in isolation.

To that end, the leadership and staffs of United States Central Command and USPACOM continue to engage in order to ensure a coordinated strategic approach that best meets U.S. interests.

Maldives, Nepal, Bangladesh, and Sri Lanka

The U.S. has extensive interests throughout the rest of South Asia. USPACOM is working to advance relations with Maldives, Nepal, Bangladesh, and Sri Lanka. Collectively, these countries are important collaborators with the U.S. for combating terrorism and transnational criminal activity.

South Asia is susceptible to natural disasters, but these nations lack the national resources and capacity to respond and therefore require significant assistance from the international community to respond to such contingencies. Toward that end, USPACOM is working with South Asian militaries to build capacity so that they may better address the challenges in their sub-region.

USPACOM is helping the Maldives address the growing threat from Somali piracy and transnational crime. In Nepal, USPACOM's focus is humanitarian assistance and efforts to support the peaceful integration of members of the Maoist People's Liberation Army into the Nepalese security forces.

Bangladesh continues to partner with the U.S. to enhance regional security, and USPACOM has initiated multiple programs to improve Bangladesh's civil-military trust, transparency, and cooperation.

Engagement with Sri Lanka is limited to humanitarian assistance until allegations of human rights abuses are resolved.

Violent Extremism

Violent extremism is among the most pervasive and urgent challenges in South Asia. VEOs are associated with a wide-range of activities, which include supporting insurgencies that seek political autonomy and fomenting conflict between nuclear-armed India and Pakistan as a means of spreading radical Islamic ideology. Consequently, USPACOM continues to expand its relationships with host-nation militaries and CT agencies to increase regional capacities to counter this threat.

Responding to VEOs requires a comprehensive and cooperative CT strategy focused on developing CT capacity with India and other South Asian partners, particularly regarding the threat posed by Pakistan-based Lashkar-e-Tayyiba (LeT).

LeT involvement in the November 2008 attacks on Mumbai, India, validates India's concerns regarding terrorist threats originating from outside India. Significantly, LeT deliberately targets westerners and specifically engages coalition forces in Afghanistan. USPACOM is keenly aware of the threat posed by this terrorist organization and continues to develop a coordinated multi-national and U.S. interagency approach to address this global threat.

South Asia Force Posture Assessment

USPACOM force posture in South Asia is relatively limited, even though this sub-region is significant and faces various challenges.

The only continuous U.S. force presence in the region is at Diego Garcia. These forces are primarily prepositioned ships used for rapid response but are not typically used for routine training or engagement. For the most part, U.S. military engagement in South Asia is achieved with transient and deployed forces and PACOM Augmentation Teams (PATs); however, plans are underway to expand these teams to assist host nations in their counterterrorism efforts.

OCEANIA

U.S. ally Australia and, to a lesser extent New Zealand, are the most prominent forces in Oceania. The sub-region's challenges are generally associated with numerous Pacific island countries spread across the South Pacific. In concert with other U.S. government agencies, allies, and partners, USPACOM seeks an Oceania sub-region that cooperates and collaborates to solve regional and international security challenges, maintains capable security or defense forces that abide by international norms, and effectively manages resources.

USPACOM strives to maximize Maritime Domain Awareness and information sharing, assists with the development of responsive decision-making architectures, contributes to improved maritime interdiction capabilities, and cooperates to ensure the existence of, and adherence to, appropriate legal authorities. Oceania's economies are largely reliant on fishing and other natural resources, and tourism. These Pacific island countries are also vulnerable to illegal activities such as illicit drug trade, human and weapons trafficking, smuggling of goods, illegal fishing, and other transnational crimes related to under-governed spaces within the maritime domain. Sub-region national capacity to conduct surveillance and police respective exclusive economic zones is uniformly limited. A coordinated effort to assist with sub-regional challenges is the Oceania Maritime Security Initiative. Under this program, U.S. Navy vessels already transiting the region embark U.S. Coast Guard shipriders to support maritime domain awareness and information sharing between the Coast Guard and five Pacific island countries (Republic of the Marshall Islands, Federated States of Micronesia, Palau, Tonga, and Kiribati). PACOM and the Coast Guard are currently reviewing the possibility of embarking law enforcement representatives from these five Pacific island countries. Natural disasters are frequent and challenge domestic and regional HA/DR capacities and require coordinated international response assistance. In that vein, the 2011 PACIFIC PARTNERSHIP mission will deliver assistance to the people of the Micronesia, Tonga, Vanuatu, Papua New Guinea, and Timor-Leste.

Australia

The U.S.-Australia alliance, which marks its 60th anniversary this year, comprises the most significant partnership in the Oceania sub-region, and indeed one of the most significant in the entire Asia-Pacific. Australia remains a resolute, capable ally that is firmly committed to enhancing global and regional security – it is the largest non-NATO contributor of forces to the coalition efforts in Afghanistan -- and to providing assistance throughout the Pacific. Australia is significant to the entire AOR; its stature throughout the region is expanding in multiple ways. An especially promising relationship is developing between Australia, Japan, and the U.S., which will advance multilateral security efforts throughout the region.

Australia values advancing defense cooperation with the U.S., particularly through training events and acquisition programs. This year, TALISMAN SABRE—a large-scale, biennial, combined-arms exercise that focuses on strengthening U.S.-Australia military-to-military capability—will occur with the participation of over 20,000 U.S. and Australian military personnel. U.S. and Australian militaries also collaborate extensively in areas such as information sharing; intelligence, surveillance, and reconnaissance; HA/DR; and most recently space and cyber security.

New Zealand

New Zealand is also a recognized leader in Oceania. Our two nations share many security concerns such as terrorism, maritime security, transnational crime, and proliferation of WMD. While New Zealand's nuclear policies preclude a formal resumption of our alliance, we have close defense cooperation and information sharing with New Zealand, and value New Zealand's contribution of its forces to security operations throughout the Asia-Pacific as well as Afghanistan. New Zealand remains active in security initiatives, from stabilization efforts in Timor-Leste and the Solomon Islands, to operations in Korea, Sudan, and throughout the Middle East. New Zealand has also contributed its multi-role maritime patrol vessel to regional, multilateral HA/DR operations. Additionally, the New Zealand Defense Force supports National Science Foundation efforts in Antarctica by serving as the primary staging area for the multinational Operation DEEP FREEZE.

Compact Nations

USPACOM partners with three Compact Nations: the Federated States of Micronesia, The Republic of the Marshall Islands, and the Republic of Palau. U.S. obligations to these nations under the Compacts of Free Association are met through implementation of Homeland Defense planning and preparation. The Compacts provide these nations with critical economic assistance. In return, the U.S. gains strategically important access. Within this framework of authority and responsibility, the United States is entitled to significant access privileges in these nations and retains the right to deny access to the military forces of other nations.

Oceania Force Posture Assessment

Current U.S. military force posture in Oceania is limited, but it includes facilities in the Marshall Islands, joint facilities in Australia, and operations by occasional transiting or deployed forces. Due to the sub-region's geography, U.S. force presence is largely limited to maritime forces and occasional HA/DR efforts. Australia is supportive of extensive visitation and engagement by U.S. forces and has recently indicated an interest in doing more to facilitate USPACOM regional posture. The U.S. and Australia are considering forms of access, supply support, joint training, and rotational presence.

EXERCISE AND ENGAGEMENT PROGRAMS

In order to maintain ready forces and to plan, train, and exercise to accomplish the full range of military contingencies, USPACOM requires annual Congressional support for its Joint Training and Exercise Program (JTEP). This essential program is comprised of 18 major exercises and involves joint military forces, interagency activities, and 27 of 36 USPACOM partner nations. JTEP is funded by the Combatant Command Exercise and Engagement Program Fund, more commonly known as CE2, which directly influences USPACOM's ability to conduct directed and opportunity-based Joint training exercises and theater security engagement events in the Pacific region. Thus, Congressional support for the CE2 program plays a vital role in maintaining security not only in the USPACOM AOR, but in Combatant Commands worldwide. USPACOM also supports improving and enabling realistic joint training through the Pacific Joint Training Strategy. The Joint Pacific Alaska Range Complex (JPARC) is used to conduct training and to support joint exercises and mission rehearsals.

A new interagency program is the Global Security Contingency Fund, which will provide an integrated State and Defense capacity-building tool for allies and partners. By joining resources of both Departments, the U.S. can more effectively and efficiently align priorities and integrate planning to enhance U.S. ally and partner capabilities.

MILITARY CONSTRUCTION SUSTAINMENT

MILCON is another important enabler of ready forces in the Pacific. Because of the MILCON Subcommittee's support in the past, the USPACOM MILCON program continues to meet transformation, operational readiness, and quality-of-life requirements. Military construction projects for readiness and quality of life are discussed below. The two other significant MILCON programs—DPRI and ROK transformation—are included in the previous Japan and ROK sections, respectively.

Readiness

In addition to the MILCON required to implement transformation initiatives in the Pacific, USPACOM Service Components continue efforts to sustain readiness at their major installations (i.e., those facilities with a military-civilian population of more than 2,500 personnel). For FY2012, the Service Components in the Pacific seek a total of $1.6 billion to sustain the readiness capabilities of their installations. The items included below are the significant MILCON requests for FY2012.

U.S. Army Pacific (USARPAC). USARPAC maintains ten major installations throughout the Pacific, including those in Alaska, Hawaii, ROK, and Japan. MILCON funds will help ensure USARPAC soldiers are trained and equipped to enable full spectrum operations. Thus USARPAC seeks $297 million for FY2012 readiness-related MILCON programs, of which $114 million is requested for an aviation task force hangar at Ft. Wainwright, Alaska. Joint Base Elmendorf-Richardson, Alaska, requires $78 million for phase two of an engineer brigade complex and for tactical vehicle parking. Wheeler Army Airfield in Hawaii requires $73 million to complete the first phase of a combat aviation brigade complex. Schofield Barracks in Hawaii requires a $32 million military vehicle wash facility.

U.S. Pacific Fleet (PACFLT). PACFLT manages five major naval facilities throughout the AOR and has operational forces at several installations on the West coast of the U.S. mainland. PACFLT installations provide vital forward presence for the world's largest fleet, including the homeport of the only forward-deployed aircraft carrier (in Japan). Readiness MILCON requirements total $229 million in FY2012. Two naval bases in Washington State require $160 million in support of the strategic weapons facility. California's requirement totals $77 million at the naval bases in Coronado and Ventura County for an aircraft maintenance facility and aircrew training facility, respectively. Finally, a $10 million electrical upgrade and a $7 million electronics shop conversion in Hawaii are essential for PACFLT's readiness requirements in FY2012.

U.S. Pacific Air Forces (PACAF). Nine major PACAF installations are located in the region, stretching from Alaska through Hawaii and Guam to Japan and Korea. This web of air bases provides USPACOM integrated expeditionary Air Force capabilities to conduct their many peacetime and wartime missions. The Air Force continues its plan to make Guam the hub for strike and refueling assets in the Asia-Pacific and requests $212 million for FY2012. Guam has also been building projects to establish the Pacific Regional Training Center and seeks $29 million to continue this effort.

U.S. Marine Forces Pacific (MARFORPAC). MARFORPAC manages nine major installations located in Arizona, California, Hawaii, and Japan. MARFORPAC installations in the Asia-Pacific provide the launching point for forward deployed combat ready Marine forces both ashore and afloat. MILCON projects supporting MARFORPAC readiness is expected to cost $788 million. Vehicle Maintenance, a Mountain Warfare Training Center and Land Expansion will cost $44 million in 29 Palms, California. Marine Corps Air Station Yuma, Arizona, requires $162 million for two aircraft maintenance hangars and infrastructure improvements. A total of $287 million for construction at Camp Pendleton Marine Corps Base, California, is requested toinclude projects to support the arrival of the MV-22 and a $29 million Infantry Squad battle Course Defense Range. Finally, $57 million will be used to construct an airfield operations complex at Marine Corps Base Hawaii.

Defense Logistics Agency (DLA). The vastness of the Pacific AOR increases the importance of providing and sustaining sufficient forward-based logistics infrastructure. The services and materiel that DLA provides to the Pacific is a critical enabler. DLA requests $105 million for readiness related projects. Joint Base Pearl Harbor-Hickam in Hawaii requests $14 million to upgrade refueling truck parking and to reconfigure a warehouse to operate more effectively, which will result in long-term cost savings. Eielson Air Force Base, Alaska, requests $17 million to upgrade a rail line to improve refueling capabilities. Naval Base Coronado, California, needs $22 million to replace fuel storage tanks and pipelines. The fourth increment of funding to replace a fuel storage facility at Naval Base Point Loma, California, requires $27 million.

Quality of Life

Quality of Life for active duty soldiers, sailors, airmen, Marines, and their dependents continues to be a key enabler and retention factor for these Americans serving overseas—often thousands of miles and many time zones away from family and loved ones. In FY2012, $620 million will ensure adequate housing, medical, education, family support, and recreation opportunities are provided to these men, women, and their families. The following include the significant projects:

Bachelor Housing. In FY2012, USPACOM requests $254 million for five barracks projects. Specifically, the Air Force plans to construct barracks in Osan, Korea, and Eileson Air Force Base and Joint Base Elmendorf-Richardson, Alaska, worth $165 million. The Army requests $89 million for new barracks at Camp Henry and Camp Carroll in Korea.

Family Housing. USPACOM Service components have made tremendous strides using MILCON and the Military Family Housing Privatization Initiative (MHPI) to renovate and replace antiquated housing throughout USPACOM.

While MHPI has accelerated the pace of modernization, MILCON is still required overseas in Japan and Korea since privatization is not authorized in these foreign countries.

In FY2012, USPACOM requests $113 million for family housing. Specifics for each Component are as follows: PACAF seeks $45 million to improve over 1,300 homes at Misawa and Kadena Air Bases in Japan. The Navy requests $45 million to improve 219 homes at Yokosuka, Atsugi, and Sasebo Naval Bases in Japan. MARFORPAC requests $23 million to completely renovate 76 homes at Iwakuni Air Station in Japan.

Medical. Providing outstanding medical care in high-quality facilities continues to be an important priority.

Although no medical projects are being submitted in the USPACOM AOR due to more pressing needs in other areas, several non-MILCON-level projects will be executed in FY2012. Additionally, a comprehensive medical needs assessment for all services in Hawaii is currently being conducted to better enable joint planning of future projects.

Family Support and Recreation. The Army requests $26 million in Hawaii to expand an existing fitness center and build a new pool and $18 million for a new Child Development Center for elementary school children.

The Marine Corps seeks $19 million for a multi-use recreational field and $24 million for a child development center at 29 Palms, California. The Navy asks for $47 million to build a new fitness center at Naval Base Coronado, California.

USPACOM ORGANIZATIONS

The following units uniquely contribute to the USPACOM mission:

Joint Intelligence Operations Center (JIOC)

The PACOM JIOC is the central hub for intelligence across the theater, responsible for managing intelligence requirements at the strategic level and providing operational support to Components and Subordinate Commands. The size and complexity of the Asia-Pacific demand a federated intelligence approach, leveraging the intelligence, surveillance, and reconnaissance contributions of all USPACOM intelligence enterprises. The JIOC also partners with the offices of the Director of National Intelligence, the Undersecretary of Defense for Intelligence, and national and allied intelligence agencies and centers. In addition to traditional sources and methods, open source Intelligence is an area that can provide critical insights and understanding with modest, targeted resources.

Asia-Pacific Center for Security Studies (APCSS)

APCSS supports USPACOM multi-national security cooperation and capacity-building efforts through international executive education and tailored assistance programs that educate and connect key regional security practitioners. APCSS brings together military and interagency civilian representatives to identify cooperative solutions to security challenges. These communities of interest include vice presidents, ministers of defense and foreign affairs, chiefs of defense, and ambassadors.

Joint POW/MIA Accounting Command (JPAC)

JPAC's mission is to achieve the fullest possible accounting of all Americans missing as a result of past conflicts. JPAC successfully accomplished 75 investigation and recovery operations globally last year and is projected to execute 75 investigation and recovery operations in 2011. Expanded operations will begin in FY 2012 as JPAC begins the path to the congressionally mandated 200 annual identifications. Discussions between JPAC and the governments of India and the Philippines continue in an effort to resume investigation and recovery operations in those countries. Meanwhile, operations in North Korea and Burma remain suspended.

Joint Interagency Task Force West (JIATF-West)

JIATF-West is the USPACOM Executive Agent for countering drug-related transnational crimes in the Asia-Pacific. This unique mission provides intelligence support and partner-nation capacity building. JIATF-West currently supports U.S. law enforcement agencies through Task Force initiatives focused on Iranian, Eurasian, and South Asian drug trafficking organizations that pose a regional and national security threat.

Conclusion

The preceding testimony outlines the Asia-Pacific strategic environment, addresses the optimal force posture in each of the four sub-regions, and outlines MILCON recommendations necessary to properly support USPACOM forces and their families. Among the many notable challenges are a belligerent DPRK and dangerous VEOs, which remain dangerous concerns for which the U.S. must be prepared. China's rapid military expansion coupled with its unclear intent poses a concern to the U.S. and many regional nations; however, opportunities for collaboration between the U.S. and China are also apparent. Overall the region remains ripe for peaceful growth, cooperation, and prosperity, due in large measure to mutual interests shared by our five regional U.S. treaty allies and many more security partners. Longstanding support from the American people and Congress for the first-rate resources in use by the 330,000 men and women of U.S. Pacific Command enable these forces to protect and defend the homeland while advancing U.S. interests throughout the Asia-Pacific.

In: Combatant Commands in the United States Armed Forces ISBN: 978-1-61942-924-6
Editors: Quincy N. Butler and Harris A. Simmons

Chapter 7

POSTURE STATEMENT OF GENERAL DOUGLAS M. FRASER, UNITED STATES AIR FORCE COMMANDER, UNITED STATES SOUTHERN COMMAND BEFORE THE 112TH CONGRESS SENATE ARMED SERVICES COMMITTEE*

April 5, 2011

INTRODUCTION

Chairman Levin, Senator McCain, and distinguished members of the committee: thank you for the opportunity to appear before you today to discuss United States Southern Command's accomplishments and future efforts in Central and South America and the Caribbean. Over the past year, we worked in close collaboration with U.S. government agencies and our partner nations to respond to the unprecedented natural disaster in Haiti and to the ongoing threats to regional security. This year, with the support of Congress, we will build on our accomplishments and continue to foster close cooperation and engagement throughout the region. We will also continue to evolve as a joint and interagency organization that promotes U.S. national and regional security interests through enduring partnerships.

These partnerships are not only enduring; they are essential. U.S. Southern Command envisions sustaining a shared partnership for the Americas; all nations working together to address problems of mutual concern. Under this vision, each exercise, program, and operation we conduct in the region augments the training of our joint forces, improves our ability to work with partner armed forces, and enhances the capabilities of our partners to confront regional security challenges. In addition, our programs directly integrate with and support other U.S. government agencies' efforts to enhance citizen safety, democratic governance, and economic prosperity. We also continuously coordinate our programs with other U.S.

* These remarks were delivered as testimony given on April 5, 2011. General Douglas M. Fraser, United States Air Force Commander, United States Southern Command before the 112th Congress Senate Armed Services Committee.

government departments and agencies. Our response to the January 12, 2010 earthquake in Haiti demonstrated the effectiveness of these efforts.

Before continuing, I would like to thank Congress for funding the construction of U.S. Southern Command's new headquarters in Miami. This state-of-the-art building ensures that we are fully prepared to accomplish our mission: we are ready to conduct joint and combined full-spectrum military operations and to support whole-of-government efforts to enhance regional security and cooperation. The new headquarters enhances internal and external collaboration, improves our ability to conduct operations, and raises quality of life for assigned personnel. Our integrated, interagency headquarters significantly enhances our collaborative approach in working to achieve our strategic objectives in the region.

REGIONAL CONTEXT

Positive Trends

Latin America and the Caribbean are comprised of a multitude of cultures, languages, heritages, and histories. The United States is connected to this region by more than physical proximity; increasing travel and trade ensure our countries remain connected culturally, socially, and economically.[1] We are also connected by many shared values and a commitment to democratic ideals. The majority of countries throughout the region seek to consolidate the democratic, security, and economic progress achieved in recent years. U.S. Southern Command endeavors to support our partner nations in these efforts through enduring engagement and continued collaboration.

The region's recent history is characterized by sustained economic growth that benefits Latin America, the Caribbean, and the United States. Over the past 12 years, U.S. trade with countries in the region grew at a faster rate than with China or the European Union.

Although trade with Latin American and Caribbean economies still makes up a small percentage of overall U.S. trading activity (8.3 percent of all U. S. trading activity in 2009), this share grew by 15.3 percent over the past fourteen years.[2] Economic indicators throughout the region have been generally positive: growth rates averaged 3.4 percent per year over the past decade and regional GDP grew 6 percent in 2010, [3] due in large part to strong economies in South America.

This significant economic growth has allowed some of our regional partners to invest in social and educational programs designed to alleviate poverty and reduce inequality. Both poverty and extreme poverty in the region fell by 3 percent from 2009 to 2010.[4] Income inequality in Latin America and the Caribbean is exhibiting some signs of lowering, thanks in part to targeted social investments.[5] In 2003, the Brazilian government launched "Bolsa Familia," which provides income support to poor families. In return, families commit to keeping their children in school and taking them for regular health checks.[6] As of 2008, Bolsa Familia has reached 46 million people [7] and has contributed to the improvement of income distribution in Brazil, resulting in the lowest levels of income inequality in the country's recent history.[1] [8] Chile's strong economic performance in the past decade has permitted the government to invest heavily in hospitals, housing, education and pension reform.[9] Between 1990 and 2000, poverty rates were reduced from 40% of the population to 20%; the 2009 poverty figure is currently 11.5%. [10,11]

Sustained economic growth and positive social developments have been nurtured by a strong regional commitment to democracy. In the past decade, there have been numerous free and fair national elections resulting in peaceful transfers of power.

Across the region, more than 60 percent of people surveyed prefer democratic governance to any other political system.[2] [12] Regional militaries have also made great strides improving professionalism, subordinating to civilian rule and respecting human rights.

Challenges to Regional Security

Despite these positive economic, social, and political gains, several threats to security and stability in the region remain. Natural disasters wreak havoc and create humanitarian crises; social exclusion and poverty remain pervasive; and threats to democratic consolidation persist. As you know, the Americas, our common home, is vulnerable to many forms of natural disasters: hurricanes, earthquakes, volcanoes, and floods are regular occurrences. Although the region was largely spared from the ravages of hurricanes this past year, it was devastated by two major earthquakes and experienced significant flooding. Inevitably, the region will be impacted by additional natural disasters in the coming year.

While improving in some countries, poverty remains an ongoing challenge, particularly in Central America.[13] In many countries, poverty is difficult to reduce because of restraints on social mobility due to race and social class. [14] This social stagnation creates openings for criminal organizations to recruit new members who see crime as an opportunity for socio-economic advancement. Positive change in social mobility is slowly occurring, but not at a rate that will significantly reduce the influence of criminal organizations in the short term.

While many countries in Latin America and the Caribbean continue consolidating their democracies, some governments have hollowed out democratic institutions and eroded constitutional checks and balances—the key ingredients essential for a functioning democratic system. These undemocratic measures go against the shared values of the Inter-American Democratic Charter. Though the Department of State and the U.S. Agency for International Development (USAID) are the lead agencies in supporting democracy and good governance initiatives to address such measures, U.S. Southern Command engages with the armed forces in the region to promote professionalization, respect for human rights, and subordination to democratically-elected governments.

While natural disasters, social inequalities, and undemocratic tendencies undermine regional stability, threats to citizen safety and border security represent broad concerns across the region. Weak institutions, inadequate support for the rule of law and lack of independent judiciaries limits accountability for corrupt government officials, business leaders, and criminals. In too many countries, less than 5 percent of all violent crimes are prosecuted.[15] Widespread impunity undermines state institutions and provides safe haven for Transnational Criminal Organizations (TCOs) to operate in an environment of lawlessness. An estimated $100 billion per year in illicit goods—drugs, weapons, counterfeit products—and an estimated 100,000 humans are trafficked through porous borders throughout the region. [16,17].

Violence is an inherent aspect of illicit activity; it is no coincidence that the countries in Latin America with the highest rates of violence are besieged by TCOs and criminal gangs. TCOs and supporting criminal elements exploit weak institutions and corrupt officials to conduct their illicit operations with impunity. As a result, insecurity is a fact of life for many of the citizens in the region. In recent years, the Central American corridor has seen a dramatic increase in illicit trafficking and brutal violence and is now the most violent region in the world outside of active war zones.[18] This is due, in part, to success in stemming illicit trafficking elsewhere in the region, which has pushed their activities into Central America. According to the United Nations Office on Drugs and Crime, homicides in Latin America and the Caribbean increased from 19.9 per 100,000 people in 2003 to 32.6 per 100,000 people in 2008 (the U.S. murder rate is 5.0 per 100,000 people). [3] [19] Much of the violence related to TCOs is connected to the protection of trafficking routes and internal power struggles, but the criminality associated with the illegal drug trade and TCO activities increases the level of related crimes, including kidnapping, murder, money laundering, and firearms trafficking, the majority of which go unpunished.

Confronting the threat posed by TCOs to citizen safety requires coordinated diplomatic, law enforcement, and military cooperation among countries in the region. In countering this threat, Department of Defense efforts are aligned and coordinated with our interagency partners. We focus our efforts on the *consequences* of security challenges in the region, and support our interagency and international partners in confronting the *causes* of these challenges.

U.S. Southern Command strongly supports increased Congressional funding to our interagency partners—including USAID and the Departments of State, Justice, and Homeland Security—to strengthen regional civilian law enforcement agencies and judiciaries. We will continue to improve our collaboration with international and interagency colleagues to combine our efforts to disrupt and reduce transnational threats to the United States and regional security.

Extra-Regional Actors in the Region

As the nations of Latin America and the Caribbean develop, they look beyond the hemisphere for trading partners, diplomatic support, and sources of aid, leading to increased activity in the hemisphere by various extra-regional state actors. U.S. Southern Command views this activity as both an opportunity and a challenge. We have a long history of regional security cooperation with the armed forces of countries such as the United Kingdom, the Netherlands, France, and Spain, and we also work with Canada on many issues related to hemispheric security. Additionally, we welcome activities by other countries when they are conducive to regional security and stability. However, as evident in the following observations, objectives for such activities are difficult to discern.

For example, China has engaged with Latin American militaries through high-level personnel exchanges [4] and arms sales. In summer 2010, China sold 18 K-8 light attack and training aircraft worth millions to Venezuela. Earlier significant arms transfer agreements include air surveillance radars to Venezuela and Ecuador, as well as K-8 aircraft to Bolivia. Additionally, China has become one of the largest providers of investment and trade in the region. With a large appetite for the natural resources needed to supply its manufacturing sector, China's imports of raw materials from the region reached $41 billion in 2009.[20] The region is now also the second-largest destination for Chinese investments, which extend to local manufacturing as well as resource extraction.[21,22]

Although not a significant investor in the region, Russia has also found markets for arms sales in Latin America and the Caribbean. From 2002-2009, arms transfer agreements between Russia and Latin America increased several fold, peaking at $5 billion in sales in 2009.[23] However, overall military spending remains low in Central and South America, and we expect some specific incidences of Russian arms and equipment sales to enhance the region's ability to counter TCOs [5]. In other cases, though, these sales have the potential to undermine regional stability. My principal concern with Russian arms in the region is the large number of man-portable air defence systems and automatic weapons sold to Venezuela, and the potential they could reach the hands of organizations like the Revolutionary Armed Forces of Colombia (FARC). Beyond arms sales, Russia is also participating in infrastructure development programs. For example, in 2010 Russia agreed to assist Argentina—as well as Venezuela—with nuclear energy programs. Russia is also widening its influence in the region by expanding diplomatic activities beyond its traditional allies of Cuba, Venezuela, and Nicaragua. Finally, Iran continues expanding regional ties to support its own diplomatic goal of reducing the impact of international sanctions connected with its nuclear program. While much of Iran's engagement in the region has been with Venezuela and Bolivia, it has nearly doubled the number of embassies in the region in the past decade[6] and hosted three regional heads of state in 2010.[7] Currently, Iranian engagement with Venezuela appears to be based on shared interests: avoiding international isolation; access to military and petroleum technologies; and the reduction of U.S. influence. Together with our interagency partners, U.S. Southern Command will continue to monitor Iranian activity in the region consistent with law and policy to ensure that U.S. laws and international sanctions are respected, and that our existing partnerships remain strong and well-functioning.

In addition to extra-regional state actors, members of violent extremist organizations (VEOs) from the Middle East remain active in Latin America and the Caribbean and constitute a potential threat. Hezbollah supporters continue to raise funds within the region to finance their worldwide activities. Several entities affiliated with Islamic extremism are increasing efforts to recruit adherents in the region, and we continue to monitor this situation closely. Additionally, we deploy military information and civil affairs teams to under-governed spaces to help our regional partners to hinder these recruitment efforts and counter VEO propaganda.

FOCUS AREAS

As we look to the future, United States Southern Command will continue supporting whole-of-government efforts that enhance the United States' role as an enduring partner of choice in the region. We will continue to do our part to sustain a region of secure, stable, and prosperous partner nations that work cooperatively to address shared challenges. In this regard, we will focus in three key areas: countering TCOs and illicit trafficking; providing humanitarian aid and disaster relief; and supporting peacekeeping operations.

Counter Illicit Trafficking

TCOs and the illicit trafficking they conduct continue to be the primary threat to regional security. These groups construct flexible, resilient networks which use multiple paths to support illicit activity. In countering this international threat, working with our partners, we must attack TCOs in a broad, coordinated manner, to include demand reduction; eradication and regulation of source materials; suppression of money laundering; interdiction of the illicit shipments as they transit to the United States and other end-user countries; and ultimately the disruption and dismantling of TCOs operating in the region. However, TCOs are increasingly sophisticated and have proven resilient and adaptive to attempts to disrupt their operations. They are innovative; to minimize and avoid detection and interdiction by U.S. and regional authorities, criminal organizations have begun using self-propelled fully submersibles to conduct illicit trafficking from South America to Central America and Mexico. These submersibles, built in the jungles of western Colombia and Ecuador, provide TCOs with a multi-ton, long-range cargo capacity. They are hard to detect and difficult to intercept. U.S. Southern Command is working with our interagency partners, the military services, and our partner nations' armed forces to counter this evolving threat.

Joint Interagency Task Force South (JIATF-South) in Key West, Florida is the center of U.S. maritime interdiction efforts in the Caribbean Basin and eastern Pacific. Using information from law enforcement, JIATF-South detects and monitors suspect aircraft and maritime vessels, and then provides this information to international and interagency partners who have the authority to interdict illicit shipments and arrest members of TCOs. This past year, JIATF-South and our international and interagency partners were directly responsible for interdicting 142 metric tons of cocaine, 3,419 pounds of marijuana, and 309 arrests, denying TCOs $2.8 billion in revenue.

JIATF-South's collaborative, interagency approach serves as the model for our partnerships with other combatant commands and U.S. government agencies. U.S. Southern Command works directly with U.S. Northern Command to synchronize Department of Defense operations in the Western Hemisphere, prevent TCOs from exploiting seams in our AORs, and coordinate the employment of our combined resources. Specifically, U.S. Southern Command and U.S. Northern Command are coordinating counter-TCO actions with Guatemala, Belize, and Mexico to enhance our combined efforts to reduce trafficking along their borders. In addition, U.S. Southern Command coordinates counter-TCO activities with our other partners throughout Central and South America and the Caribbean.

Active engagement with our partner nations is a key component to effectively counter transnational criminal activities. U.S. Southern Command supports U.S. government security initiatives in Colombia, Central America, and the Caribbean. Our ongoing strategic partnership with Colombia—undertaken within the framework of the Colombian Strategic Development Initiative (CSDI) originally developed by the U.S. Embassy in Bogota—serves as a model for integrated collaboration. CSDI aligns the U.S. government support to Colombia with the Colombian government's National Consolidation Plan, a whole-of-government effort to expand state presence and services in targeted areas where poverty, violence, illicit crop cultivation, and drug trafficking have historically converged. Colombia has suffered from decades of violence and instability as narco-terrorist groups, financing their activities through drug trafficking, waged an insurgency against the government. While challenges remain, the security situation today in Colombia is drastically different, thanks in large part to the sustained efforts of the Government of Colombia, supported by Plan Colombia and its corresponding U.S. government-sponsored initiatives. Since August 2002, more than 54,000 combatants from Colombia's illegal armed groups have demobilized. Of these, 58 percent demobilized collectively as a result of an agreement between the Government of Colombia and the paramilitary United Self-Defense Forces of Colombia (AUC). The FARC's strength has declined from over 18,000 in 2002 to around 8,000 today and its territorial control has decreased significantly. The successes of Plan Colombia and Colombia's own democratic security initiatives resulted in the acceleration of economic development; security and stability have helped Colombia achieve annual growth rates averaging 4 percent. [24]

U.S. Southern Command's role in supporting the execution of Plan Colombia and its corresponding programs and initiatives involved equipping and training the Colombian armed forces; the sharing of technical expertise; and the facilitation of technology transfers. The armed forces of Colombia continue to yield positive results. In 2010, with the support of the U.S. Embassy country team and U.S. Southern Command, the Colombian armed forces planned and executed a string of operational successes, including eliminating Victor Julio Suarez Rojas, also known as "Mono Jojoy" (the FARC's long-time military leader and fourth in command) and 15 other high-ranking FARC members. I ask for continued Congressional support for Colombia; your support has resulted in a valuable and reliable partner directly involved in countering illicit trafficking and promoting regional stability. As noted in the 2010 National Drug Control Strategy, while Colombia's gains have been impressive, they are reversible, and we value continued Congressional support to CSDI and other initiatives. [25]

Our focus on countering transnational criminal organizations and their illicit trafficking activities extends beyond Colombia to include Central America and the Caribbean. In support of the U.S. government's two sub-regional initiatives to improve citizen safety—the Central American Regional Security Initiative (CARSI) and the Caribbean Basin Security Initiative (CBSI)—U.S. Southern Command will continue to support interagency efforts to interdict illicit trafficking in international waters and airspace. Through Theater Security Cooperation activities, we will continue to enhance the capacity and capability of our partner militaries to operate within their respective territories and to support bilateral and multilateral counter illicit trafficking operations.

With its porous borders, lack of surveillance capabilities, and under-governed areas, Central America has become the TCOs' preferred transit zone to the United States. TCOs support and use a spectrum of destabilizing activities to conduct their operations, to include corruption, intimidation, extortion, kidnapping, targeted violence, and terror tactics. Confronting this spectrum requires a sophisticated, orchestrated strategy that both guides efforts to meet current challenges as well as sets a framework for disrupting future TCO adaptations.

Our current plan to counter the trafficking threat in Central America is to support U.S. interagency efforts and help build self-sustaining regional military capacity to increase the cost and consequences to TCOs of using the Central American transit zone. Under initiatives like Enduring Friendship, we facilitated the procurement of maritime interdiction assets and command, control, and communications (C3) capabilities for Central American and Caribbean Basin countries[8]. To strengthen international borders, we are facilitating technology transfers that support Department of Homeland Security training that is improving our partners' ability to detect and interdict illicit shipments at international crossings. We are also providing training and equipment to partner nations' ground forces to strengthen their capacity to respond to TCOrelated events requiring a military response.

However, the limited capabilities of Central American states have allowed Mexican TCOs to establish convenient points of entry for illegal drugs coming from South America. Nearly all cocaine destined for the U.S. crosses the Guatemala-Mexico border. The expansion of Mexican TCOs into Central America has created even more violence and crime, and a significant decline in citizen safety. Focusing specifically on this vulnerable Mexico-Guatemala-Belize border area, we are engaged in planning with our U.S. Northern Command, interagency, and partner nation colleagues, and are on solid footing towards developing a regional operations capability among these three countries.

With Caribbean Basin countries, we want to reduce the ability of TCOs to expand their operations. To achieve this goal, we will support CBSI and leverage existing regional initiatives in the Caribbean Basin like the Caribbean Community and the Regional Security System in the Eastern Caribbean to build capacity to conduct aerial and maritime surveillance and interdict illicit trafficking.

Humanitarian Assistance/Disaster Relief (HA/DR)

HA/DR consists of two separate but complementary missions. Humanitarian assistance provides support for basic human needs—food, water, shelter, and sanitation—to populations temporarily or chronically underserved. Disaster relief reduces the human suffering associated with natural disasters which cause the disruption of normal transportation and commerce and destroy infrastructure. Our annual humanitarian and civic assistance exercises provide valuable training for U.S. military medical, engineering, and combat support personnel, while complementing the Department of State and USAID's goal of advancing community development and hemispheric prosperity. Disaster relief activities go beyond deploying our own forces when disaster strikes. We also seek to improve our partner nations' capacity to conduct disaster relief operations within their own borders, and when possible, outside their borders. We envision a region in which mutual assistance is the norm.

Our annual training exercises in the Caribbean Basin and Central America help improve our ability to conduct disaster relief and humanitarian assistance at both the tactical and operational levels. In 2010, we conducted 76 medical readiness training exercises (MEDRETEs), resulting in the treatment of 276,827 patients throughout the region. During our annual engineering exercises—NEW HORIZONS and BEYOND THE HORIZONS—our forces built or renovated ten schools, six health centers, six sanitation facilities, two police stations, and seven water wells in communities in Nicaragua, Panama, Guatemala, the Dominican Republic, and Haiti. These exercises also helped increase response capabilities of the participating partner nations; during the course of the training, our forces provided valuable training to first responders and disaster managers in the host countries.

Joint Task Force Bravo (JTF-B) at Soto Cano Air Base in Honduras provides regional support for responding to natural disasters and supporting counter drug operations with our partners in Central America and the Caribbean. In 2010, JTF-B medical personnel conducted four Medical Capability Projects in El Salvador, Guatemala, Honduras, and Nicaragua, treating 6,981 patients and also supported relief efforts in Haiti and in Guatemala after the eruption of the Pacava volcano and the landfall of Tropical Storm Agatha. I thank Congress for its continued support of JTF-B, especially for the appropriation of funds to support construction of new barracks at Soto Cano.

Unquestionably, the most significant 2010 HA/DR operation for U. S. Southern Command was Operation UNIFIED RESPONSE in Haiti. The situation after the January 12th earthquake was grim: over 230,000 people killed, 300,000 wounded, and one million people displaced in our hemisphere's poorest country. Critical infrastructure, including the Port-au-Prince airport and seaport, was destroyed or unserviceable. Thirteen of fifteen government ministries were destroyed, crippling the Haitian government's ability to respond.

We established Joint Task Force-Haiti in support of USAID, the lead federal agency for the disaster response effort. Our response was immediate: within hours the airport was re-opened; within days maritime transportation was reestablished, allowing the influx of food, water, and medical supplies. U.S. soldiers, sailors, airmen, and marines provided critical rescue, medical, and relief supply distribution support for the Haitian people. During Operation UNIFIED RESPONSE, U.S. forces delivered 2.3 million meals, 17 million pounds of bulk food, 2.6 million bottles of water, and almost 150,000 pounds of medical supplies. In addition, Department of Defense medical personnel—operating on the ground as well as onboard USNS COMFORT—treated almost 10,000 patients, and conducted 1,025 surgeries and 343 medical evacuations. In partnership with non-governmental organizations and the private sector, U.S. Southern Command coordinated both the delivery of additional relief supplies worth $36.2 million and the integration of 200 civilian medical specialists and translators into our relief efforts.

JTF-Haiti completed its mission on May 15th, but the end of our operation did not signify the end of our support to Haiti. Instead, we transitioned to a smaller mission consisting of targeted humanitarian and civic assistance exercises. As part of NEW HORIZONS HAITI 2010, approximately 500 personnel—mainly from the Louisiana National Guard—deployed to Haiti from June to September 2010. Engineers completed thirteen projects, building schools, improving wells, and constructing sanitation facilities. Medical forces conducted ten MEDRETES and established clinics that each served four to five thousand patients. Equally noteworthy, the exercise involved forces from another country in the region. Belize—partnering with the Louisiana National Guard under the State Partnership Program—deployed an engineer company to assist with reconstruction efforts. This example of regional collaboration represents the type of capability we actively seek to build and sustain throughout Latin America and the Caribbean. Our commitment to Haiti is ongoing. As Haiti rebuilds, U.S. Southern Command will continue to conduct annual humanitarian and civil assistance exercises and respond to lingering effects of the devastating earthquake, as well as other humanitarian challenges.

U.S. Southern Command also supported U.S. government disaster relief efforts in Chile following a devastating 8.8 magnitude earthquake on February 27th 2010. In support of USAID, approximately 150 U. S. military personnel deployed to Chile and worked with local and international responders to deliver 300,000 pounds of relief supplies. Working with Chilean counterparts, a U.S. Air Force Expeditionary Medical Support team treated more than 300 patients. U.S. Southern Command also partnered with the private sector, which donated $1.2 million in transportation assistance to ship 40,000 meals-ready-to-eat to Chile. It is important to note that Chile possessed the internal capacity to effectively respond to the disaster, requiring limited assistance from the U.S and other nations. This epitomizes the type of capability we seek to promote with our other regional partners.

Finally, during hurricane season in the Caribbean Basin, U.S. Southern Command remains prepared to assist any partner nation in the region affected by a tropical cyclone. We design our annual maritime deployment, CONTINUING PROMISE, to conduct humanitarian assistance in the Caribbean Basin while maintaining readiness to respond to disaster relief efforts, if requested. This past year, USS IWO JIMA—with a Special Purpose Marine Air Ground Task Force of approximately 500 Marines, 128 NGO personnel, and 44 partner nation personnel embarked—deployed to the Caribbean Basin between July and November. During eight port visits, U.S. medical personnel treated 45,517 patients and performed 329 surgeries; dental personnel treated 15,472 patients; and veterinarians treated 26,969 animals. Engineers completed 23 projects over the course of the deployment. In late October, when Hurricane Tomas was forecast to strike Haiti with Category 3 strength, U.S. Southern Command diverted USS IWO JIMA from a scheduled port visit in Suriname to a safe location near Hispaniola as part of U.S. government response preparations. Following landfall of the storm, USS IWO JIMA moved into the area quickly and U.S. Marine helicopters conducted several damage assessment flights. Fortunately, Hurricane Tomas only struck a glancing blow to southwestern Haiti, and relief organizations already on the ground were able to respond to the minimal damage caused by the hurricane.

Peacekeeping Operations (PKO)

Our partner nations exhibit a tremendous capacity to conduct peacekeeping operations in the region and around the globe. During Operation UNIFIED RESPONSE, U.S. Southern Command drew on the strengths of our South American regional partners who comprised the United Nations Stabilization Mission in Haiti (MINUSTAH). Led by a Brazilian general officer, peacekeepers from Argentina, Brazil, Chile, Bolivia, and Uruguay were on the ground when the earthquake occurred and were instrumental in maintaining security and leading the initial response. MINUSTAH continues to play an invaluable role in Haiti through the on-going efforts of our hemispheric partners. U.S. Southern Command also supports peacekeeping efforts through our partnership in the Department of State's Global Peace Operations Initiative (GPOI). Joining with nine countries in the region, we develop or enhance national training capabilities and equip potential peacekeeping units for deployment of U.N. Peace Support Operations.[9] We conduct the annual PKO Americas exercise, which is designed to improve partner nation capacity to plan and conduct peacekeeping operations. In the coming year, we will continue to support GPOI and our regional partners who take the lead in peacekeeping operations.

ENDURING ENGAGEMENT

Military-to-Military

Building partner nation capacity and enhancing interoperability is at the core of everything we do in our AOR. In implementing this strategy, we facilitate exchanges, seminars, and training exercises throughout the region with our partner militaries.

A cornerstone of our engagement strategy is the International Military Education and Training program, which provides professional development for foreign military officers and senior enlisted personnel from Latin America and the Caribbean. Each year, U.S. Southern Command helps send approximately 5,000 students from the AOR to attend U.S. military training programs across the Department of Defense, to include the Western Hemisphere Institute for Security Cooperation, the Center for Hemispheric Defense Studies, the Inter-American Defense College, and the Inter-American Air Forces Academy. Our goal is to encourage our partner nation militaries to promote institutional professionalism within the context of democratic governance.

These programs are vital to building and sustaining relations with our partners throughout the region.

Our training and education programs also promote respect for human rights with our military partners. These programs remain important across the region, but hold particular relevance in the few countries whose militaries are being asked by their governments to assist local police forces in supporting and enhancing internal security. U.S. Southern Command's programs support our military partners in preparing to assume these roles and help them provide clear instructions to their soldiers to support and respect human rights. Our human rights programs and initiatives remain an important aspect of our engagement throughout the region.

Our engagement strategy is reinforced through our Foreign Military Interaction (FMI) exercise program. Every year U.S. Southern Command sponsors seven military exercises specifically designed to facilitate interoperability, build capabilities, and provide venues to share best practices among the military and security forces in the region. Our largest multinationalexercise, PANAMAX 2010, brought together eighteen nations from the Western Hemisphere to train for the defense of the Panama Canal [10]. Other key FMI exercises in the region include TRADEWINDS, FUERZAS COMANDO, and UNITAS. These exercises provide a venue for participating militaries to train together and maintain security and stability within the region.

Interagency

Very few threats in the region require a conventional military response; as a result, the predominant security challenges we face are best addressed through the coordinated efforts of many U.S. government agencies.

U.S. Southern Command headquarters is organized to support this coordination and collaboration; 27 representatives from 12 different agencies are embedded throughout our structure.

This integration is both efficient and effective, allowing us to combine resources, perspectives, and expertise to collectively address issues in the region.

Our interagency partners contribute to the development of strategic plans and participate in our joint exercises and operations, a cooperation that is critical to our success in the region. We continue to seek innovative ways to orchestrate our efforts across the U.S. government to maximize our results.

REQUIREMENTS

In order to successfully achieve our strategic objectives in the region, U.S. Southern Command has identified requirements in two key areas: *Intelligence, Surveillance, and Reconnaissance* and *Foreign Military Sales.*

Intelligence, Surveillance, and Reconnaissance (ISR) are critical enablers of U.S. Southern Command's operations. Effective countering of illicit trafficking operations is contingent upon our ability to detect and monitor illegal activities. As demonstrated during Operation UNIFIED RESPONSE, ISR is also valuable for supporting HA/DR operations. ISR components—such as improved imagery intelligence, wide area coverage, sensor integration, signals intelligence, moving target indicators, layered ISR architecture and management tools, and biometrics—will improve our ability to synthesize a common operating picture to better support our operations in the region.

An additional source of valuable regional insight is information available and disseminated on the Internet. Social media and social networks provide opportunities for increased regional awareness and improved collaboration with our partners. U.S. Southern Command is improving our ability to analyze social media sources such as Twitter and blogs so we can identify regional trends early and accurately. When appropriate, we are also seeking to improve and expand the use of the All Partners Access Network, an online community that promotes collaboration among governmental and non-governmental organizations during exercises and operations.

We continue to work with the defense industry and the Defense Advanced Research Projects Agency to identify promising technologies that match our requirements. Specific needs include: flexible, persistent manned and unmanned aerial vehicles; light detection and ranging technologies for foliage penetration; fast and flexible unmanned surface craft to support maritime domain awareness; acoustic and electronic sensor technologies to detect semi- and fully-submersible craft; commercial satellite radars with the ability to detect high-speed watercraft; next generation Over-the-Horizon radars; non-electro-optical imagery which enables change detection; and the associated Tasking, Collection, Processing, and Dissemination architecture. Individually and collectively, all of these technologies enable our operations and represent opportunities to develop regional capabilities.

Foreign Military Sales (FMS) and Foreign Military Financing (FMF) are key components in the security assistance the U.S. government provides our partner nations. The goal of these programs is to increase partner nation capability and capacity to help us address threats to security and stability. Ideally, FMS would be an efficient process to rapidly support and enhance partner nation capabilities and deliver products that are tailored and appropriate for a nation's requirement. Unfortunately, the current program is inflexible and does not allow for efficient coordination within the interagency community. An improved FMS program would increase interoperability, strengthen military and economic ties, and maximize the efficient use of resources. The Defense Security Cooperation Agency has begun reforms that represent a promising start to addressing the inadequacies of the current FMS system. We also support efforts to pool State and Defense resources for the purpose of funding more robust and comprehensive security sector assistance programs to respond to emergent challenges and opportunities. These and other improvements are necessary to effectively develop a comprehensive, integrated security assistance program.

Conclusion

U.S. Southern Command is committed to being a trusted, reliable partner of choice in the region. Our success will depend on our ability to engage effectively and transparently with regional militaries, partner nation governments, and our interagency partners. We actively work with the countries in our AOR to build enduring, mutually beneficial partnerships that address our shared security concerns: violence and instability caused by TCOs and illicit trafficking, and the repercussions of natural disasters. Each training exercise and operation in the region is designed to increase partner nation capabilities to help us confront these challenges; we envision a region that is capable and willing to share the responsibility of hemispheric security and stability. Although we have experienced successes, we remain vigilant for evolving threats; watchful for new opportunities; and willing to engage with our partners to enhance our international, interagency, and public-private relationships. None of the progress we made this year would be possible without the dedication and hard work of our military and civilian personnel, the support of their families, and the cooperation by the men and women from our partner agencies who serve alongside us. I thank Congress again for your continued support to all the dedicated professionals at U.S. Southern Command as we serve together to accomplish our mission.

Douglas M. Fraser
General, U. S. Air Force

References

[1] U. S. Department of Commerce, Office of Travel and Tourism Industries. *2009 Market Profile: Central America. 2009 Market Profile: South America. 2009 Market Profile: Caribbean.* and *2009 United States Resident Travel Abroad.* From 2002-2009, travel from the AOR to the U. S. increased by 32 percent and from the U. S. to the AOR by 41 percent.

[2] Hornbeck, J.F. (2010). *U.S. Latin America Trade: Recent Trends and Policy Issues.* Congressional Research Services: June 25, 2010.

[3] The Economic Commission for Latin America (2010). *Preliminary Overview of the Economies of Latin America and the Caribbean 2010.* Available online at: http://www.eclac.org/cgi-bin/getProd.asp?xml=/prensa/noticias/comunicados/9/41979 /P41979. xml&xsl=/prensa/tpl-i/p6f.xsl&base=/tpl-i/top-bottom.xsl

[4] The Economic Commission for Latin America (2010). *Social Panorama of Latin America 2010.* Santiago, Chile: November 2010.

[5] United Nations Development Programme. *2010 Human Development Report.* Available online at: http://hdr.undp.org/en/media/ HDR_2010_ EN_Complete_reprint.pdf

[6] International Labour Office (2009). *Bolsa Familia in Brazil: Context, Concept and Impact.* Geneva: 2009.

[7] MDS (Ministerio do Desenvolvimento Social e Combate à Fome) (2009). *Programas de Transferencia Condicionada de Renda: o caso do Bolsa Familia.* Presentation made

in the Employment and Social Policies Committee of the ILO Governing Body, March (Geneva). Cited in ILO Report (referenced above).

[8] Instituto de Pesquisa Económica Aplicada. 2008. *PNAD 2007: Primeiras Análises:Pobreza e mudança social.* Comunicado da Presidencia. N0 9, Sept. (Brasilia, IPEA).

[9] World Bank (2004). Shanghai Poverty Conference: Case Study Summary (Chile). Available online at: http://info.worldbank.org/ etools/docs/ reducing poverty/case/24/ summary/Chile%20Summary.pdf

[10] The Economic Commission for Latin America (2010). *Social Panorama of Latin America 2010.* Santiago, Chile: November 2010.

[11] Ibid

[12] Latinobarómetro Corporation (2010). *Latinobarometro Annual Report: 2010.* Santiago,Chile. Available online at: www.latinobarometro.org

[13] Council on Foreign Relations (2008). *Latin America and the United States: New Directions for a New Reality.* Independent Task Force Report No. 60.

[14] Azevedo, V. (2009). Social mobility in Latin America: a review of existing evidence. Interamerican Development Bank: Washington, D.C.

[15] *Reaching the untouchables.* The Economist. March 11, 2010. http://www. economist.com/node/15663302?story_id=15663302

[16] xvi *The Globalization of Crime: A Transnational Crime Threat Assessment.* U. N. Office of Drug Control. 2010.

[17] Congressional Research Service, *Trafficking in Persons in Latin America and the Caribbean*, October 16, 2009

[18] Casas-Zamora, Kevin (2010). *Paying Attention to Central Americas Drug Trafficking Crisis.* The Brookings Institute: October 27, 2010. Available online: http://www. brookings.edu/opinions/2010/ 1027_central_america_ drugs_casaszamora. aspx

[19] xix United Nations Office on Drugs and Crime (2007). *Crime and Development in Central America: Caught in the Crossfire.* Geneva: May 2007; International Crisis Group (2008). *Latin America Drugs I: Losing the Fight.* Latin America Report No. 25, March 14, 2008.

[20] Arnson, C. and Davidow, J., eds. (2011). *China, Latin America, and the U.S.: The New Triangle.* Woodrow Wilson International Center for Scholars. Washington, D.C.

[21] Pomfret, J. (2010). *China invests heavily in Brazil, elsewhere in pursuit of political heft.* Washington Post: July 26, 2010, section A01.

[22] Associated Foreign Press. *China investment in Peru tops in Latin America: Official.* April 22, 2010.

[23] Bancroft-Hinchey, T. (2010). *Russian Arms Sales Overtake U.S. in Latin America.* Published September 9, 2010. Available online: http://english. pravda.ru/russia/ economics/09-02-2010/112127-russian_arms_sales-0/#

[24] *Country Data: Colombia.* World Bank Database. Available online at http://data. worldbank.org.

[25] Office of National Drug Control Policy (2010). *2010 National Drug Control Strategy.* Washington, D.C.: page 83.

End Notes

[1] Brazil's 2010 Gini coefficient=0.55

[2] According to the Latinobarometro 2010: regionally, 15 percent of respondents support an authoritarian system; 23 percent did not know or were indifferent. In comparison, the USAID-sponsored AmericasBarometer survey by the Latin American Public Opinion Project indicates 70 percent support for democracy in countries surveyed in 2010 (www.lapopsurveys.org).

[3] Comparing homicide rates puts the magnitude and deterioration of citizen insecurity in Latin America into perspective: in 2010, the homicide rate in Afghanistan was 8.6 per 100,000; 13.7 in Iraq; 71.0 in El Salvador; and 77.0 in Honduras. Sources (multiple): United Nations Mission to Afghanistan; Iraq Body Count; and National Civilian Police figures.

[4] At least 12 in 2010

5 For example, Peru's purchase of 8 Russian-made helicopters in July 2010 to conduct counter illicit trafficking operations.

[6] Iran had six embassies in the AOR in 2005 and ten in 2010.

[7] The Presidents of Bolivia, Guyana, and Venezuela.

[8] Enduring Friendship countries include: Dominican Republic, Bahamas (funding provided in FY06 when they were in the U. S. Southern Command AOR), Jamaica, Panama, Honduras, Nicaragua, El Salvador, Costa Rica, and Belize. Enduring Friendship was expanded into the Eastern Caribbean under the "Secure Seas" rubric providing funding to the Regional Security System HQ, Barbados, Grenada, Antigua & Barbuda, St Kitts & Nevis, Dominica, St Vincent& the Grenadines, St Lucia, Trinidad & Tobago, Suriname, and Guyana.

[9] GPOI countries: Belize, Nicaragua, Honduras, Guatemala, El Salvador, Peru, Paraguay, Uruguay, and Dominican Republic. Chile, Colombia, and Ecuador are pending diplomatic agreements.

[10] PANAMAX participants: Argentina, Belize, Brazil, Canada, Chile, Colombia, Dominican Republic, Ecuador, El Salvador, Guatemala, Honduras, Mexico, Nicaragua, Panama, Paraguay, Peru, and Uruguay.

In: Combatant Commands in the United States Armed Forces ISBN: 978-1-61942-924-6
Editors: Quincy N. Butler and Harris A. Simmons

Chapter 8

POSTURE STATEMENT OF ADMIRAL ERIC T. OLSON, USN COMMANDER UNITED STATES SPECIAL OPERATIONS COMMAND BEFORE THE 112TH CONGRESS SENATE ARMED SERVICES COMMITTEE*

March 1, 2011

Mr. Chairman and distinguished members of the committee, thank you for this opportunity to provide an update on the United States Special Operations Command (USSOCOM). Our Special Operations Forces give us much cause for great pride and it is my deep privilege to represent them to you, and especially to do so for the fourth time as their commander.

My intent today is to describe the current status, activities and requirements of Special Operations Forces. I'll begin by briefly describing USSOCOM and its assigned Special Operations Forces.

As many of you know, USSOCOM is a creation of Congress, legislated into being in 1986. A relatively small number of Army, Navy and Air Forces units designated as Special Operations Forces were assigned to USSOCOM, with Marine Corps forces joining the Command just over five years ago.

Before the establishment of USSOCOM, the Nation's Special Operations Forces had generally not been treated as a top priority. They now thrive under the focused attention of a single headquarters and a dedicated budget. In the 24 years since USSOCOM was established, Special Operations Forces have repeatedly proven their value, often under extraordinarily demanding conditions.

In many ways, USSOCOM is a microcosm of the Department of Defense, with ground, air and maritime components, a global presence, and authorities and responsibilities that mirror the military departments, military services and defense agencies. We take pride in the diversity of our people and our mission.

* These remarks were delivered as testimony given on March 1, 2011. Admiral Eric T. Olson, USN Commander United States Special Operations Command before the 112th Congress Senate Armed Services Committee.

One of our headquarter's functions is to synchronize Department of Defense planning against terrorists and their networks globally. This is complex work that connects us across DoD and into other U.S. government departments and other nations' military forces. The effects of this are manifested in a series of planning documents that guide specific actions by the Services and Combatant Commands.

Primarily, USSOCOM organizes, trains and equips Special Operations Forces and provides those forces to the Geographic Combatant Commanders under whose operational control they serve. The Command also develops special operations strategy, doctrine and procedures for SOF employment and develops and procures specialized equipment for the force. Our key subordinate commands are U.S. Army Special Operations Command, Naval Special Warfare Command, U.S. Air Force Special Operations Command, Marine Corps Forces Special Operations Command and Joint Special Operations Command. Within these commands are the legendary Special Forces or Green Berets, SEALs, Air Commandos, Rangers, Night Stalker helicopter crews, and the modern version of yesterday's Marine Raiders. Our force also includes the active duty practitioners of Civil Affairs Operations and Military Information Support Operations, and all of the instructors, logisticians, administrators, analysts, planners, communicators, doctrine writers and other specialists who are key to our ability to meet our Nation's needs. Most are active duty military, but we depend heavily on our Guard and Reserve units and the government civilians and contractors who perform duties that don't require a uniformed service member.

We now total close to 60,000 people, about a third of whom are career members of Special Operations Forces, meaning those who have been selected, trained and qualified to earn the Military Occupational Specialty or skill code identifier of a SOF operator.

The activities of the force are as varied as its character. From high-risk, high-intensity counterterrorist raids; to meticulous intelligence analysis; to providing first response during a natural disaster; to launching from submerged submarines; to training and accompanying foreign counterparts; to working with local leaders to determine what will bring value to their village; to providing supporting precision fires to fighting troops from orbiting aircraft – SOF personnel are in vital roles, in key places, performing essential tasks.

Our presence is generally small and agile, inherently joint and persistent. Our formations normally include an array of attached capabilities that are necessary to optimize the force – including female Cultural Support Teams, Tactical Air Controllers, Military Working Dogs, interpreters, maintenance and repair personnel, Explosive Ordnance Disposal technicians and others. SOF rarely dominate an area with their mass, so they must work with indigenous forces and the local civilian population to accomplish their missions. This is often complicated, demanding and high-risk.

Each of the Geographic Combatant Commanders who will appear before you is well served by the Special Operations Forces that are deployed to his region, although the balance is heavily weighted towards U.S. Central Command. In fact, about 85 percent of deployed SOF are directly engaged in Operations NEW DAWN and ENDURING FREEDOM. I will defer to the regional commanders to highlight the contributions of Special Operations Forces in their theaters. I will just say here that, although the precision counterterrorism missions certainly receive the most attention, SOF are conducting a wide range of activities in dozens of countries around the world on any given day – at the request of the host government, with the approval of the U.S. Ambassador and under the operational control of the U.S. Geographic Combatant Commander.

To support these forces and activities, USSOCOM invested in many specialized programs and equipment. As the commander responsible for the preparation and readiness of SOF, I focus on developing and sustaining operational skills and capabilities, training and maintaining the quality of the force, caring for its families, and ensuring that our people have the right equipment in sufficient quantity. I also carefully monitor global military and political trends in my role as the senior advisor on the employment of SOF.

Among USSOCOM's most important functions is the management of Major Force Program–11 (MFP-11). MFP-11 is provided to the Commander of USSOCOM to address requirements that are "SOFpeculiar" in nature, and it is the essential fuel that enables Special Operations Forces to meet the Nation's needs. It provides for the conduct of advanced and unique training, the timely and flexible fielding of equipment, and the capability to rapidly and effectively project our force. In Fiscal Year 2012, the request for MFP-11 funds totals $10.5 billion in baseline and Overseas Contingency Operations (OCO) funding. This is an increase of 7 percent over the FY11 request, and every dollar is necessary to meet the ever-increasing demands placed on our Special Operations Forces.

At the forefront of budget discussions is the acknowledgment that many of the current expenditures funded by OCO are, in fact, part of USSOCOM's baseline requirement in the "new normal." This was highlighted by the department last year when a commitment was made to eventually move funding required to execute Overseas Contingency Operations into the baseline as part of the SECDEF's initiative to "rebalance" the force. However, USSOCOM will continue to rely on OCO funding over the next few years as the phased transfer to the base budget occurs. For example, in the FY 2012 budget submission thirty-four percent of the total MFP-11 request is OCO funding. For some higher intensity SOF elements, the OCO percentage is greater than 75 percent. USSOCOM will carefully prioritize and manage the OCO to base transition. Overall, we are in a fiscally satisfactory condition, but the force requires continued support. The president's Budget Request for FY12, if approved, is an essential step towards meeting the growing demand on our force by providing USSOCOM the resources required to sustain critical programs and initiatives. Now, I would like to highlight some of these key efforts.

PROGRAMS

USSOCOM continues to expand and recapitalize its rotary and fixed wing aviation fleets. This year we began modification of the last of the originally planned 61 MH-47G helicopters, while starting procurement of eight additional MH-47Gs. We are also fielding the first of 72 planned MH-60M helicopters as part of our recapitalization of MH-60 K/L platforms. The tilt-rotor CV-22, having demonstrated its capabilities on multiple deployments, must remain on plan to ensure enhanced future mobility capabilities for SOF. USSOCOM's MC-130Ws, rapidly modified with a Precision Strike Package utilizing SOF's Joint Acquisition Task Force, are providing armed overwatch and mobility to deployed SOF as an interim augmentation to our Vietnam-era AC-130 gunship fleet. We are on a path to ultimately recapitalize the gunships with AC-130J models. The MC-130J program is on track to replace our aging MC-130Es and MC-130Ps. Our Non-Standard Aviation Program is delivering a variety of smaller aircraft to provide intra-theater airlift capacity and we continue to grow our aviation foreign training capability in support of the Geographic Combatant Commanders' engagement plans.

USSOCOM is also modernizing its maritime mobility systems. We will award competitive prototype contracts later this year for Combatant Craft – Medium (CCM) as replacements for the Naval Special Warfare Rigid Hull Inflatable Boat (RHIB). We have realigned resources from the Advanced SEAL Delivery System (ASDS) and the Joint Multi-Mission Submersible (JMMS) to fund the development of a family of Dry Submersibles as part of our undersea mobility strategy. These will be launched from surface ships or Dry Deck Shelter-equipped submarines. As part of this modernization program, we will explore expansive and flexible approaches that are supportive of the Secretary of Defense's intent to streamline acquisition processes and accelerate delivery times.

Special Operations Forces continue to rely on a wide range of ground mobility vehicles, often leveraging Service and Department investments. Modified to meet the wide variety of SOF mission sets and provide enhanced crew protection, vehicles such as the MRAP have been essential to SOF teams operating in dispersed and rugged terrain throughout the USCENTCOM Area Of Responsibility. These vehicles, as well as our other ground mobility systems, will remain relevant well into the future as we synchronize our long-term sustainment strategy with the Services.

We continue to invest in airborne manned and unmanned Intelligence, Surveillance and Reconnaissance (ISR) programs, relying heavily on the Services to expand capabilities and capacity that benefit DoD across the board. USSOCOM is moving toward a relatively small number of manned and unmanned ISR systems; essential Processing, Exploitation, and Dissemination (PED) capabilities; and supporting communications architectures. One of the most noteworthy improvements within special operations over the last few years has been the growth of advanced communications and networking capabilities through our expeditionary SOF Information Enterprise (SIE). As our portion of the Department's Global Information Grid, the SIE provides network independence while maintaining connectivity into the global interface, and links Special Operations Forces across the globe into a common network. This connectivity shortens the decision cycle for SOF operators worldwide and allows more rapid information sharing. The research and rapid development of these types of technologies is an inherent strength of special operations.

USSOCOM, inherently joint in all it does, is in a unique position to leverage and apply Service and Department Science and Technology (S&T) efforts to rapidly field new technologies on the battlefield. USSOCOM's "Rapid Exploitation of Innovative Technologies for SOF" (REITS) program, enables innovative new capabilities to be developed and inserted quickly into the battlefield - advanced "talk and jam" capabilities for SOF vehicles; mobile repair and maintenance "shops in a box"; to solar panel energy technology that supports SOF in remote locations. USSOCOM also seeks to expand its biomedical research and development activities. To date, USSOCOM has pushed "state of the art" combat medicine with modest resources through the Tactical Combat Casualty Care program. However, we also have great need to explore innovative methods of treating our wounded members so that they may be reintegrated and returned to duty as rapidly as possible.

As a force that operates from the tropics to the Arctic regions, from under water to high elevations, and from peaceful areas to violent combat zones, Special Operations Forces serve as an ideal "control group" for Service R&D investments that can result in significant benefits across DoD.

USSOCOM's development of the Joint Acquisition Task Force (JATF) concept enabled accelerated acquisition and fielding of urgent SOF capabilities. First demonstrated on the MC-130W Dragon Spear program, USSOCOM expanded use of the JATF concept to address many emerging requirements of SOF warfighters. Innovative approaches such as the JATF, coupled with a professionally trained and certified SOF acquisition corps that stays in close and frequent contact with the operators, continue to ensure that USSOCOM remains as a vanguard of rapid acquisition within DoD.

USSOCOM's acquisition planning, collaboration, and continuing dialogue with the Services continues to improve as we become more efficiently effective while rapidly moving capabilities to the warfighter.USSOCOM, in conjunction with USD AT&L, initiated a series of Acquisition Summits with the Military Department Acquisition Executives to minimize programmatic disconnects and to better align requirements, co-sponsorship opportunities, funding efficiencies, and contracting actions among MFP-11 programs and Service-related/dependent programs. These periodic meetings offer a level of transparency among all our accounts that enables us to seek common solutions for Service-wide requirements and to better invest in SOF-peculiar modifications or special capabilities. This forum identified several opportunities, which if supported by Congress, would enable more efficient execution of SOF unique acquisitions.

USSOCOM is making a significant investment in Military Construction (MILCON) to address shortfalls resulting from fielding new capabilities, a growing force structure and aging infrastructure that was inherited without a future recapitalization budget. To address the shortfall, the Command's 2012 budget submission is based on a MILCON roadmap that identifies over 300 prioritized requirements valued at more than $5 billion between 2012 and 2025. Specifically, our FY12 budget submission includes 33 of these projects, valued at $631M across eight states and representing 9 percent of the Command's projected base budget request – a near record level. This investment demonstrates a commitment to addressing our critical infrastructure needs. To continue this effort, the Command's new Strategic Planning and Programming Guidance raised the MILCON funding minimum from 4 to 6 percent to support this priority in future budgets.

A Congressional action that enhanced the effectiveness of our force is our Section 1208 authority. This authority to reallocate limited MFP-11 funds remains a key tool used by widely dispersed SOF to leverage indigenous forces in support of counterterrorism operations. USSOCOM is appreciative of the increase to $45 million provided by Congress in the FY11 National Defense Authorization Act, as it provides us the ability to support ongoing operations with a measure of flexibility should a contingency arise. Continuation of Section 1208 authority provides enhanced effectiveness to our force both strategically and tactically.

Initiatives Our primary challenge is the need to carefully manage the growth of Special Operations Forces, even in these periods of high demand, in order to ensure the continued quality the nation expects. I have stated in my last three posture hearings that SOF's organic manpower growth should be in the range of 3-5 percent per year. That is the pace we have sustained to great effect over the past several years and our FY12 budget submission continues this pace. But 3-5 percent growth within USSOCOM will not answer the increasing demand for our force unless it is matched by the Military Services' commitment to attach supporting and enabling forces at a commensurate rate. SOF units must include a limited amount of these enabling forces to ensure rapid response to emerging requirements, but we were designed and intended to rely on the Services to meet most of our combat support and combat service support requirements. In order to establish a predictable demand signal for these Service-provided capabilities, USSOCOM is proposing changes to the way we build, train, deploy and sustain a fully enabled force.

To better build the SOF team, we are developing a force generation system that engages the existing Service systems. In 2011, USSOCOM will strive to create a SOF Force Generation system that will be synchronized with the Services, matching their capabilities with our Special Operations core units in time to provide fully optimized force packages to the Geographic Combatant Commanders. For elements organic to SOF, such as our Civil Affairs and Military Information Support Operations, we will expand their capacities to meet the increasing demand for their capabilities.

Another challenge we face is how to effectively prepare and train the force to achieve enhanced interoperability with the General Purpose Forces (GPF). Currently in the USCENTCOM AOR, SOF is executing the counterterrorist strike mission and the Village Stability Operations mission; two of the primary lines of operation underpinning the Afghanistan strategy. SOF's key role in both is creating opportunities for enhanced interoperability with the GPF such as the deployment of the 1ST Battalion, 16TH Infantry, now assigned to the Combined Joint Special Operations Task Force – Afghanistan (CJSOTF-A); a sea change in SOF-GPF relations. Currently, we are developing initiatives that will increase inter-operational effectiveness prior to the deployment phase of the operation.

In 2011, we will continue to review and coordinate changes to Service personnel policies to further incentivize language pay for key languages such as Pashto, Dari and Arabic. We will work to develop courses of action that allow SOF reliable and predictable access to Service resources such as training ranges for our ground and aviation elements. The shortage of readily available, local ranges currently hampers SOF's ability to meet deployment training timelines and causes our operators to "travel to train," further increasing their already excessive time away from home.

Understanding the operational context of the environments in which we operate is a hallmark of SOF. Developing this knowledge and experience within our force, and understanding the value of "micro-regional" expertise allows SOF to conduct its activities with more predictable outcomes. While immersion opportunities enhance our regional sophistication, our training can never develop the level of nuanced understanding possessed by indigenous populations. To gain this high level of cultural knowledge, USSOCOM will continue to strongly support DoD's Military Accessions Vital to the National Interest (MAVNI) and the Army's Intermediate and Advanced Language Programs (IALP) to recruit and access the requisite expertise provided by native speakers. Additionally, our attached female Cultural Support Teams (CSTs) allow us to reach key elements of the population in some environments which was not previously possible. This concept of attaching females to SOF units is effective and long overdue; we are urging the services to recognize the capabilities of CSTs as essential military skills.

Finally, our efforts to become more innovative include studying the best practices of other organizations. For example, we are inspired by the ability of the World War II's Office of Strategic Services to rapidly recruit specialized talent, develop and acquire new technologies and conduct effective global operations within the period of its relatively brief existence.

To further our engagement with our international allies and partners, and within the U.S. interagency community, USSOCOM will continue to expand the Special Operations Liaison

Officer (SOLO) and Special Operations Support Team (SOST) programs. Both of these outreach efforts provide SOF experts to support and enhance their host organizations while serving as SOF liaisons. Our priority is to assign SOLO officers wherever a foreign partner has, or is planning to establish, a USSOCOM-like headquarters.

Joint operations and special operations are two growing trends in many of our partner nation military forces. One manifestation is the recent establishment of the NATO SOF Headquarters. In accordance with the 2010 National Defense Authorization Act, the Secretary of Defense designated USSOCOM as the lead component for this Headquarters – a role we will embrace and expand in an effort to advise and assist an interoperable network of global SOF.

Importantly, we remain committed to caring for our service members and their families. I am concerned about the effects of nine years of focus on combat operations on the well-being of our extended special operations community. To support the wounded and injured and their caregivers, the Command remains committed to our Special Operations Care Coalition and the Tactical Human Optimization, Rapid Rehabilitation and Reconditioning (THOR3) Program. Both programs are focused on long term care, rehabilitation and reintegration of our warriors. In an additional effort to be predictive and preventive, I established a "Pressure on the Force" Task Force to survey and analyze the effects of repetitive combat deployments over nearly a decade. Necessarily relying on soft data, collective experiences and commanders' instincts, it will try to determine what initiatives might help ease the strain and contribute to long term retention and force stability. I expect to receive the recommendations from this team within 90 days.

In conclusion, I will reinforce what I believe are the top challenges to the Command. As the Chairman of the Joint Chiefs of Staff recently stated, "...the first forces in are typically Special Forces. And the last ones out are going to be Special Forces." As we expect to remain the force of first choice for many military operations, USSOCOM must:

1. Carefully and deliberately meet the ever-increasing demand for Special Operations Forces.
2. Improve and expand our tactical and operational level skills, equipment and systems.
3. Preserve our proposed budget levels and authorities.
4. Find better structures and processes to obtain Service-provided capabilities.
5. Continue to improve our acquisition speed and agility.
6. Better understand the people and conditions in the places we go, whether to assist or fight.
7. As our most solemn duty, look after the health and wellbeing of this magnificent force from whom we ask so much.

Today's Special Operations Forces are the most capable, best prepared SOF in history. Their ingenuity, perseverance, spirit and skill continue to inspire and amaze. In significant ways, they have emerged from the shadows to make visible and dramatic impacts of great magnitude. It is my honor to have served within SOF for the last 37 years and to represent this extraordinary force today before this committee.

As always, our success is only possible because of your continued support and advocacy. Your approval of the president's Budget Request will help ensure our continued ability to address some of our nation's most daunting security challenges.

In: Combatant Commands in the United States Armed Forces ISBN: 978-1-61942-924-6
Editors: Quincy N. Butler and Harris A. Simmons

Chapter 9

STATEMENT OF GENERAL C. ROBERT KEHLER COMMANDER UNITED STATES STRATEGIC COMMAND BEFORE THE SENATE COMMITTEE ON ARMED SERVICES*

29 March 2011

Chairman Levin, Senator McCain, and members of the committee, thank you for the opportunity to present my views on United States Strategic Command's (USSTRATCOM) missions and priorities. Today marks my first appearance before you as the Commander of USSTRATCOM. I have occupied this position for a short time, but I have been involved with the Command's missions for much of my career. I was privileged to spend two years as the Deputy Commander, and most recently I served as one of USSTRATCOM's Service component commanders. USSTRATCOM's active duty, reserve, and civilian members, who are standing watch this very minute at locations across the country and around the globe, exemplify the best of today's joint force. I am privileged to lead this remarkable team, and I look forward to working with you to assure our nation's security.

America's strategic forces proudly continue their long-standing role as the foundation of our national security posture. The president of the United States has assigned USSTRATCOM the responsibility to detect, deter, prevent, and defeat attacks against the United States, its territories, possessions and bases, and to employ appropriate force to defend the nation should deterrence fail. The Command's specific mission responsibilities include planning, synchronizing, advocating, and employing capabilities to meet the nation's strategic deterrence, space operations, cyberspace operations, information operations (IO), global strike, missile defense, intelligence, surveillance, reconnaissance (ISR), and combating weapons of mass destruction (CWMD) objectives. We conduct these activities in close coordination with other combatant commands around the world. Today, I would like to

* These remarks were delivered as testimony given on 29 March 2011. General C. Robert Kehler, Commander, United States Strategic Command, before the Senate Committee on Armed Services.

describe the strategic context in which we operate and USSTRATCOM's priorities for addressing our many challenges.

STRATEGIC CONTEXT

The national security landscape continues to be marked by protracted conflict, constant change, and enormous complexity. While war remains a difficult struggle between human beings, today's operating environment is significantly different than those we experienced in the past. The number and type of actors (state, non-state, terrorist, criminal) are rapidly changing, and the distinction between combatants and non-combatants is less clear. Friend and foe alike can span global distances in seconds through space and cyberspace, and technological advances allow adversaries to cross traditional geographic and military boundaries with ease. Adversaries seek advantages by using asymmetric means to find and exploit our vulnerabilities and to defeat our advanced capabilities in air, sea, space, and cyberspace. At the same time, these adversaries wield hybrid combinations of capabilities, strategies, and tactics and operate in the shadows to present us with ambiguous indications and situations. Rapid technological evolution and the wide civil availability of formerly advanced military capabilities have also reduced —entry costs,|| making available completely new weapons and enabling actors to access capabilities that would not have been available to them in the past without significant investment. Indeed, surprise may be our deadliest foe, because it can make our plans ineffective, our training irrelevant, and, therefore, our organizations vulnerable.

The need to foster strategic stability and deter strategic conflict, ensure uninterrupted capabilities from and access to space and cyberspace, respond to traditional and non-traditional threats, and deal with surprise in an era of rapid technological advances presents USSTRATCOM with significant challenges.

Of the threats we face, weapons of mass destruction clearly represent the greatest threat to the American people, particularly when pursued or possessed by violent extremists or state proliferators. The potential of nuclear uncertainties in unstable regions adds special significance to this concern.

At the same time, today's fiscal environment will pose additional challenges regarding the means and manner with which we address the difficult global, strategic landscape. Last year, Secretary of Defense Robert Gates challenged us to foster an efficient "culture of saving" throughout the Department of Defense (DoD).

The resulting review emphasized our responsibility to maximize both mission effectiveness and taxpayer value. USSTRATCOM's exhaustive assessment of our missions identified some functions that we could reduce, consolidate with other DoD organizations, or eliminate in favor of higher priority operational requirements. We are now evaluating these initiatives with the DoD leadership and will realign resources as directed at the conclusion of this assessment.

In summary, the challenges are great, the choices are hard, and there is no textbook solution.

PRIORITIES

The 21st-century security environment demands fast, comprehensive awareness, strategic thinking, flexible planning, decentralized execution, rapid innovation, and an unprecedented emphasis on sharing information. In this environment, USSTRATCOM has been uniquely organized and positioned to shape and employ global capabilities to deter, enable, and, when needed, join with the other combatant commands to fight and win the ever changing joint fight.

First and foremost, we must guarantee a safe, secure, effective, and ready nuclear deterrent force. As affirmed by the 2010 Nuclear Posture Review (NPR), sustaining and modernizing the nuclear weapons complex, the triad of nuclear forces, the human capital, and key supporting command/control/ communications (C3) and intelligence/surveillance/ reconnaissance (ISR) capabilities is essential to retain confidence in the deterrent's long-term credibility, provide tools to combat proliferation, and assure our scientific and innovation edge.

Next, in full partnership with the other combatant commands, we must improve our plans, procedures, and capabilities to address trans-regional problems. Ongoing operations demand our full commitment, and USSTRATCOM's activities both enable and support joint operations around the world. The Command's work to synchronize and advocate for missile defense, ISR, electronic warfare, and combating WMD plans and capabilities helps bring unity of effort and flexible capabilities to trans-regional operations. Whether providing space-based communications or position, navigation, and timing (PNT) information, rapidly transmitting data around the world, or ensuring tested, capable missile defenses or other globally significant capabilities are developed, positioned, and optimally managed, USSTRATCOM is instrumental in winning today's dynamic joint fight.

Finally, we must continue to improve our capabilities and operating concepts in the important civil and national security areas of space and cyberspace. Ensuring uninterrupted access to space and space-based capabilities, improving our awareness of objects and activities in space, integrating their effects with all operational phases, improving space access, protection, and resilience, and expanding our planning and implementation for partnership operations requires that we continue our investment and that we demand acquisition results. For cyberspace, we must enhance network protection and mature our organizations, capabilities, workforce, and partnerships to ensure effective operations.

STRATEGIC DETERRENCE

In today's complex security environment, the concept of strategic deterrence must encompass strategies to deter adversaries and dissuade competitors across the full range of their capabilities. We must consider actors and capabilities in aggregate, not in a vacuum, a need that highlights the importance of a better understanding of adversaries' values, motivators, capabilities, intentions, and decision-making processes. Not every potential adversary has or seeks nuclear weapons, and modern deterrence requires broad coordination, tailored strategies, effective capabilities, international cooperation, and focused capabilities like conventional prompt global strike.

Still, USSTRATCOM's first priority is to deter nuclear attack on the United States, our allies, and our partners. Last year, the Quadrennial Defense Review (QDR), the NPR, and the New Strategic Arms Reduction Treaty (New START) discussions produced an important national consensus that affirmed the necessity of the United States' nuclear deterrent and the funding required to sustain it. The president has pledged that the United States will maintain a safe, secure, and effective nuclear deterrent as long as nuclear weapons exist. USSTRATCOM is now committed to implementing New START and to advocating for planned investment in the deterrent force. The updated "1251 Report" submitted in February of this year outlines both DoD and Department of Energy nuclear funding requirements through Fiscal Year (FY) 2021. While budget estimates will be refined as major program baselines evolve, these important investments must begin immediately. Congress approved the first of these increases for the National Nuclear Security Administration (NNSA) in initial FY 2011 continuing resolutions, and I strongly urge full funding in FY 2011 and 2012.

Nuclear Enterprise. The FY 2011 and 2012 budget requests reverse several years of downward trend in nuclear enterprise funding. These budgets provide investments in the facilities, equipment, and personnel dedicated to sustaining and managing the nation's nuclear weapons, as well as to dismantling weapons no longer needed. To emphasize the importance of this investment and to better understand the conditions, urgent needs, and impending challenges across the complex, I plan to visit each lab and production facility during my first 60 to 90 days in command.

The men and women of America's nuclear weapons complex perform uniquely difficult, highly technical, and demanding work. As our stockpile ages well beyond each weapon's originally designed lifespan, robust stockpile surveillance and assessment programs will enable strategic deterrence and stability at New START force levels. Weapon safety, reliability, and performance may change in ways we cannot fully predict, and surveillance activities permit confidence and continued stockpile certification without nuclear testing. Dedicated surveillance and life extension studies constitute the best means of informing the President and the Congress of our nuclear weapons' health, status, and requirements. The NPR's case-by-case approach to studying and selecting from the full range of life extension options (refurbishment, reuse, and, if needed, replacement) ensures the best future for our stockpile.

Today, a narrow window is available to synchronize weapon sustainment efforts for the W76-1 and B61 (full scope) life extension activities—cost-effectively introducing improved safety and security features, avoiding a second B61 nuclear refurbishment in the 2020s, and potentially reducing the stockpile by consolidating four legacy B61 variants into a single weapon. In addition, a Nuclear Weapons Council study of W78 ICBM and W88 SLBM life extension program options will examine opportunities to use modular fuze components and develop a possible common warhead, potentially reducing costs and supporting long-term capability sustainment. These and future actions that evaluate ways to reduce warhead numbers and types through stockpile commonality and flexibility offer the opportunity to continue accomplishing our strategic deterrence mission while also achieving the goal of a smaller, more efficient stockpile.

Strategic Delivery Vehicles. The NPR also affirmed the continuing need for the nuclear triad, which provides the president with multiple options for a variety of scenarios. The value of the triad lies in its flexibility and responsiveness to the changing world environment and in its ability to hedge against technical failure, geopolitical change, or a breakthrough in another nation's capabilities. America's strategic forces require continued investment to ensure their future capability, and USSTRATCOM is actively engaged with our Service partners to define and advocate for necessary nuclear force modernization and recapitalization programs.

ICBMs. The widely dispersed and responsive Minuteman III ICBM force provides high readiness, low operating costs, and sovereign basing with multiple aim points that complicate adversary targeting. The Air Force is successfully concluding decade-long efforts to enhance safety and security and to sustain the Minuteman force through 2020. The Air Force is also evaluating requirements to sustain the force through 2030. USSTRATCOM supports these programs and is working with the Air Force on a Capabilities Based Assessment (CBA) and pre-Analysis of Alternatives (AoA) activities that begin to define options for a follow-on land-based strategic deterrent beyond 2030.

SLBMs. Ohio-class SSBNs provide an assured and highly survivable response capability, and the highly accurate Trident II D5 strategic weapon system continues to exceed the demanding operational reliability standards established almost 30 years ago. By the time they begin to retire in 2027, the Ohio-class SSBNs will have served for more than 40 years. The Navy completed an Ohio-class follow on platform AoA and, with USSTRATCOM, continues to refine specific replacement requirements. USSTRATCOM fully supports Navy efforts to maintain the current fleet, fund the necessary research and development for its replacement, and sustain the Trident II D5 ballistic missile and associated infrastructure to satisfy future deterrent requirements. For example, current infrastructure at Naval Base Kitsap-Bangor, WA lacks sufficient Explosive Handling Wharf (EHW) capacity to meet growing missile handling requirements. A second Pacific EHW wharf at Naval Base Kitsap-Bangor, WA is essential to long-term SSBN readiness.

Bombers. America's B-2s and B-52s ensure that the president has visible and flexible conventional and nuclear global strike and deterrence options. Affirming their critical deterrent role, the nuclear-capable bomber force transitioned to USSTRATCOM's day-to-day operational control in 2010. USSTRATCOM now has a far stronger voice in balancing this unique, dual-capable nuclear and conventional bomber force's day-to-day readiness, training, and operational employment. While the Air Force continues to sustain mission-critical systems, it will also soon begin developing a new long-range, dual-capable penetrating bomber. Coupled with the development of a new bomber, two additional capabilities will ensure the viability of the air-breathing leg of the Triad for decades to come. Air Force investments will sustain the Air Launched Cruise Missile through 2030 (or until a suitable replacement is fielded), ensuring standoff capability for the long term. Further, the bomber force must be supported by a fleet of new aerial refueling tankers to extend their range and assure the bombers' strategic and extended deterrence roles. USSTRATCOM supports Air Force progress toward ensuring the long term health of the airborne component of our strategic capability.

Nuclear Command, Control, and Communications. A reliable, assured C3 capability from the President to the nuclear forces is fundamental to an effective strategic deterrent. National leaders, commanders, ISR assets, and strategic forces must share assured linkages to confidently understand and effectively address nuclear mission demands. Current systems require investments to ensure reliability and address looming capability gaps in our National Leadership Command Capability.

A new Strategic Command and Control Complex and Nuclear C3 node at Offutt Air Force Base, Nebraska is at the center of our nuclear C3 plans. The FY 2012 Budget seeks a first increment of $150M to begin replacing the aging and fragile Curtis E. LeMay building and colocated facilities. Today's building, command center, and computer systems took shape long before the IT revolution and now lack the capacity to support current mission demands. The buildings' systems strain to support numerous computer and communication systems, and the spaces occasionally experience serious heating and cooling problems, electrical failures, and other outages. For example, in December 2010 and January 2011, two water pipe ruptures caused significant system outages and dislocated staff for several days, although the Command remained capable of performing its missions due to extraordinary workarounds and the remarkable efforts of the dedicated staff and a small army of outside emergency help.

Prior to defining the current requirement, USSTRATCOM—in consultation with the Army Corps of Engineers and the Air Force—evaluated sustaining the status quo, renovating the existing facility, or engaging in new construction. The evaluation concluded that new construction offered the most operationally efficient solution to support USSTRATCOM's missions, operations, and nuclear C3 needs. The new facility will ensure an EMP-protected, flexible, sustainable, reliable, and collaborative environment with an infrastructure that meets the security challenges of today and tomorrow.

Conventional Prompt Global Strike. A limited, credible, conventional Prompt Global Strike (PGS) capability would provide the president with an important deterrent option in some strategic scenarios. Today, we still lack the ability to rapidly deliver conventional effects against fleeting or geographically isolated targets, allowing a potential adversary to establish a sanctuary using mobility and strategic depth. Research, development, test, and evaluation projects continue making progress, and I ask you to continue supporting these PGS efforts.

International Engagement: Deterring and dissuading nuclear threats in today's national security environment also requires careful attention to international relationships. While the specter of global nuclear war may be more remote than decades ago, the possibility for miscalculation between nuclear-armed states remains a perilous threat to global security. As noted in the NPR, —Enduring alliances and broad-based political relationships are the foundation of strategic stability and security.|| Indeed, many nuclear-armed states are important partners in combating proliferation. New START lowers the maximum number of U.S. and Russian strategic offensive arms, restores an important, confidence-building verification regime, and provides opportunities to continue military-to-military engagement. China's willingness to consider and study Secretary Gates' proposal for a strategic security dialogue represents an important avenue for growth between our two militaries in this area as well. USSTRATCOM will continue to support DoD, Department of State, and geographic combatant command activities to develop stable and cooperative relations with other responsible nuclear powers and will be prepared to provide advice on other arms control measures that could encompass a greater range of weapons.

SPACE

Throughout the 20th century, the U.S. and other countries developed and exploited the space domain's extraordinary potential, including changing how we navigate, communicate, and understand our world. However, the domain is increasingly congested, contested, and competitive. Guaranteeing mission assurance through adequate Space Situational Awareness (SSA), resilience, and critical-asset protection is essential. The new National Space Policy, signed by the President, and the National Security Space Strategy (NSSS), co-signed by the Secretary of Defense and Director of National Intelligence, emphasize the need to continue developing resilient capabilities which will improve our ability to satisfy combatant commanders' requirements for uninterrupted ISR, expanded military satellite communications, and PNT support. Implementing the NSSS will position the national security space enterprise to shape and strengthen the space domain's safety, stability, and security; to maintain and enhance U.S. advantages in space; to energize the U.S. industrial base by engaging a broad range of partners; to prevent and deter aggression; and to improve sustainability, acquisition, and flexibility of U.S. space capabilities.

Situational Awareness. Space Situational Awareness (SSA) is central to mission assurance and increasingly important. As part of its SSA mission, USSTRATCOM now tracks more than 22,000 orbiting objects. Approximately 1,100 of these objects are active satellites, but the remaining debris litter a variety of orbits and threatens both critical systems and human spaceflight. While space surveillance is improving, we do not yet have robust, assured, and real-time situational awareness of the orbital domain. Current and future investments should expand data integration, sharing, and exploitation; improve object detection, identification, and tracking; and advance our ability to characterize potential collisions (conjunctions). Notably, the proposed Space Fence promises to expand detection capacity more than tenfold from just two or three locations outside the continental United States and to construct a more comprehensive orbital picture. Increasing the number of objects tracked will be largely useless, however, without corresponding improvements in data integration and exploitation technologies. As part of its SSA mission, the Joint Space Operations Center (JSpOC) must also be prepared to identify and attribute purposeful space system interference and provide timely recommendations to address the interference. Without space situational awareness of the orbital domain, link segment, and supporting ground infrastructure, any plans for resilience, mission assurance, augmentation, and reconstitution will have a weak underpinning. USSTRATCOM fully supports funding for both the JSpOC Mission System (JMS) and planning and design work for a modern JSpOC facility that will facilitate a generational leap from static displays to automated, real-time visual conjunction analyses—improving our ability to protect critical space-based assets and maintain our free access to and use of space. In addition, technology will soon allow us to link multiple sensors together in a single network that will meet the needs of many users.

Cooperation. As a global domain, space and space-based capabilities operate irrespective of geographic or military boundaries. As more nations join the space-faring ranks each year and the number of objects in earth orbit grows, the need to establish norms of behavior and to improve the cooperation and collaboration among responsible space users grows as well. Our objective is to sustain a safe, stable, and secure space domain while maintaining the national security advantages space systems provide. U.S. efforts to share SSA data represent an important step toward greater international space cooperation, which should eventually help to integrate sensors and data from allies and partners worldwide and ultimately move towards a combined space operations center.

Today, the USSTRATCOM SSA sharing community includes more than 41,000 users in 141 countries. Our efforts promote the safe and responsible use of space by providing satellite operators with highly accurate predictions of close approaches between space objects for every satellite operator. Since the Secretary of Defense delegated his authority to enter into agreements with commercial entities to the USSTRATCOM Commander last September, we have concluded 19 agreements and are processing others. Each partner and each agreement signifies an operational relationship that can yield important exchanges, perhaps someday leading to a broad, international partnership for space situational awareness. USSTRATCOM fully supports expanded planning and implementation for space partnership operations among allies, coalition partners, and commercial interests and will work with our partners in the DoD and elsewhere to help review proposals to establish normalized behavior.

Space-Based Capabilities. Enabling better situational awareness will improve the overall U.S. space posture; however, long-term, uninterrupted capability from space requires equal dedication to protection, resilience, augmentation, and reconstitution of assets in space, supported by timely design and development, cost-effective acquisition, and high-confidence space launch.

Today's operating forces rely on space capabilities throughout the kill chain and beyond. Putting already stressed space capabilities that allow the joint force to navigate, communicate, see the battlefield, and strike under all conditions in the kill chain places those same valuable capabilities on any potential adversary's target list. USSTRATCOM fully supports DoD efforts to improve resilience and increase the protection of key space assets.

Launch. Reliable space capabilities also require an assured ride to orbit. Evolved Expendable Launch Vehicles (EELVs) are the DoD's primary launch vehicles and the sole U.S. vehicles for much of the national security manifest.

USSTRATCOM supports further Air Force investments in this and other programs that will assure our access to space. Additionally, improvements in manifest and scheduling processes and investments designed to sustain and ensure national launch facilities' availability for future demand will maximize synergies between launch management and national priorities.

Industrial Base Concerns. Beneath our national security space requirements lies the need for a stable, responsive, and innovative national industrial base. Since the space age began, we have rarely been so reliant on so few industrial suppliers. Many struggle to remain competitive as demand for highly specialized components and existing export controls reduce their customers to a niche government market. Careful interagency planning that more tightly defines and oversees requirements, supported by stable budgets and production rates will help sustain a national industrial base essential to commercial users, military space, and the strategic deterrent. The retirement of the Space Shuttle and other changes at NASA also injected significant concern into the solid rocket motor industrial base—an industry we cannot afford to lose. Substantial weakening of this capability would impede current strategic system sustainment and follow-on development. While industry adjustments are inevitable, DoD is developing a plan in consultation with NASA to sustain the solid rocket motor industrial base to ensure we retain right-sized, cost-efficient, and viable design, development, and production capabilities.

USSTRATCOM supports these important DoD efforts to improve program stability, increase the quantity and quality of the acquisition workforce, strengthen clarity and articulation in the requirements process, and stimulate scientific and technological advancements.

CYBERSPACE

Last fall in *Foreign Affairs*, Deputy Secretary of Defense William Lynn noted that, "Every day, U.S. military and civilian networks are probed thousands of times and scanned millions of times." Like space, cyberspace capabilities have rapidly become critical but also increasingly vulnerable. Cyberspace's pervasive presence, high importance, difficulty of attribution, and low cost of entry highlight some of our challenges. Combined with a growing, global reliance on cyberspace and its hosted capabilities, this constant evolution challenges mission assurance efforts—particularly as the threat moves from exploitation to disruption. Ensuring reliable, sustainable networks, freedom of access, and freedom of maneuver is not just a DoD problem. This is a national security problem. Assuring access demands sustained, resilient, and flexible approaches to maturing our defense capabilities, our capacity, and our cooperative relationships within and beyond the U.S. government.

Capabilities. The most important asset any commander can have is robust, up-to-date situational awareness. Cyberspace is dynamic, and specific threats require specific countermeasures. The Maginot Line failed because it was static and the defense failed to anticipate and address technological and tactical changes. "After the fact" detection and attribution don't work in cyberspace today either. The offense always has a strong advantage, overwhelming, subverting, or defeating static defenses. Continued advances in system and organization teamwork, coupled with the development and deployment of information-based capabilities and intelligence-driven sensors that "see" intrusions and can respond at equivalent speed is essential. Driven by strong, capable organizations, dynamic, agile, and informed capabilities that comprehend the network and mitigate threats at the boundary will significantly strengthen defense of DoD networks.

In response to the growing threat, last year the DoD established U.S. Cyber Command (USCYBERCOM) at Fort Meade, MD as a sub-unified command to USSTRATCOM. USSTRATCOM delegated responsibilities to USCYBERCOM to coordinate, plan, synchronize, and execute cyberspace operations in order to better defend DoD networks and to support other combatant commanders. We must accelerate the acquisition of comprehensive, shared cyber awareness tools to expand opportunities to secure critical information, reduce points of vulnerability, and develop responses to ensure warfighter access to essential information systems.

Capacity. Today, operators at USCYBERCOM and its subordinate service components work to defend against and attribute numerous information network intrusion attempts. The cyber workforce is growing, but our organizations and capabilities must also grow to keep pace with ongoing operations. USSTRATCOM is working with USCYBERCOM to improve the cyber awareness of every DoD member with access to an information system, strengthen organizations, resolve roles/responsibilities, expand partnerships, build technological and human capacity for full-spectrum cyberspace operations, and integrate cyber capabilities into every commander's plans and operations. Recruiting adequately trained and equipped cyber warriors is challenging, but fortunately young Americans grow up learning and adapting to new technological platforms from a young age. Service cyber career paths are still being developed, and these critical, technical skills need both time to develop and sustained investment to prevent their atrophy. Sustained force development emphasis and investment is essential. The U.S. is also home to the world's premier educational and commercial information technology entities. We must continue to capitalize on this capacity and partner with these organizations on our requirements and to spur domestic math and science interest. Doing so will help develop, expand, and sustain a base of cyber expertise and adapt DoD personnel processes to attract, develop, and retain the cyber professionals necessary to protect critical DoD infrastructure and preserve U.S. freedom of action in cyberspace.

Cooperation. Cyber defense must include a wide range of partners. After all, this is truly a national security issue, making interagency and allied partner engagement and information sharing essential to a robust defense. Military operations depend on the broader U.S. information technology infrastructure, and defending military networks will net fewer benefits if the wider civilian infrastructure remains at much greater risk. The Department of Homeland Security (DHS) is ultimately responsible for coordinating the protection of the —.gov|| and domestic —.com|| domains, but DoD has much to offer in terms of intelligence and technical support. The DoD – DHS Memorandum of Understanding signed last fall lays important groundwork for enhanced cooperation, mutual support, and synchronized operations.

WINNING TODAY'S FIGHT

In strategic deterrence, space, and cyberspace, USSTRATCOM both operates forces and supports the full range of military operations. The broad scope of our responsibilities and trans-regional capabilities is clearly woven into the fabric of today's operations. Winning the fight, whether we are either a supported command or are supporting the geographic combatant commands, is something our team strives to do each and every day. However, USSTRATCOM also has responsibilities to integrate, synchronize, and advocate for other capabilities with trans-regional impact, and we are dedicated to partnering with other combatant commands to improve the warfighting effectiveness of these capabilities.

INFORMATION OPERATIONS

Consistent with our mission to improve strategic joint capabilities, USSTRATCOM participated in a 2010 Secretary of Defense directed Strategic Communication (SC) and IO Front-End Assessment, designed to evaluate and recommend improvements for DoD roles, missions, definition, management, and resources for SC and IO.

As a result of the assessment, USSTRATCOM will reorganize the Joint Information Operations Warfare Center (JIOWC) at Lackland AFB, TX. Existing JIOWC resources and missions not specific to electronic warfare will be realigned to the Joint Staff, and USSTRATCOM will remain the DoD lead for Electronic Warfare (EW).

Electronic warfare. The electromagnetic spectrum spans almost every modern technological convenience. While operational plans normally assume unfettered spectrum access, this assumption is not assured.

Changing industry standards, global growth of civilian devices, military bandwidth requirements, and disruptive or destructive adversary electronic warfare capabilities all threaten to pinch or sever the shrinking electromagnetic links between national security platforms and the operating forces that rely on them.

Recognizing future threats, potential limitations, urgent warfighter needs, and the need for unified DoD advocacy, JIOWC completed several Joint Requirements Oversight Council (JROC) tasks to examine capability gaps and solutions for emerging electromagnetic spectrum threats. The National Defense Authorization Act for Fiscal Year 2010 required DoD to develop an EW strategy, submitted to Congress last year. That EW strategy concluded that we must move beyond the traditional understanding of EW by combining it with other kinetic or non-kinetic capabilities to increase U.S. combat effectiveness and achieve electromagnetic spectrum superiority. USSTRATCOM is planning to establish a Joint Electronic Warfare Center to advocate for and support DoD Joint EW capability requirements, resources, strategy, doctrine, planning, training, and operational support.

MISSILE DEFENSE

The Ballistic Missile Defense System (BMDS) exists to meet combatant commands' theater defense needs and to provide for the limited defense of the United States. Working with geographic combatant commands and the Missile Defense Agency (MDA), our efforts focus on building tailored, regional missile defense architectures using the concept of a Phased Adaptive Approach (PAA) and on meeting urgent warfighter capability needs. USSTRATCOM's work provides a comprehensive assessment of the fielded BMDS's suitability and effectiveness and combines warfighter needs for air, cruise missile, and ballistic missile defense capabilities to inform programmatic actions and guide future R&D investment priorities.

At the 2010 Lisbon Summit, North Atlantic Treaty Organization NATO allies affirmed the PAA for missile defense as a means to address the continued qualitative and quantitative growth of global ballistic missile programs. The allies also invited the Russian Federation to participate in missile defense cooperation. As a strategy, PAA applies to several geographic combatant commands, and USSTRATCOM's current challenge is to make sound, analytically-based recommendations to balance limited BMD assets worldwide. The European PAA's four phases of increasing capability are designed to defend against existing and near-term threats posed by short- and medium-range ballistic missiles and to build up defenses against long-range ballistic threats over time as those threats mature. As stated during the New START debate, the U.S. will not agree to any ballistic missile defense limitations or constraints and indeed intends to continue developing and deploying systems consistent with U.S. interests. The U.S. missile defense program is not designed to counter the strategic forces of Russia or China, but rather to address limited ballistic missile threats such as those posed by Iran and North Korea.

As various regional PAAs develop, USSTRATCOM will continually re-evaluate the standing Global Integrated Missile Defense Concept of Operations and other acquisition, deployment, basing, and employment plans for missile defense capabilities between and across all areas of responsibility. Our analysis will ensure that the joint warfighters' requirements receive deliberate management and readiness structures to ensure timely, flexible deployment, employment and redeployment of tested, understood BMD capabilities during and after crises.

Consistent with the Ballistic Missile Defense Review, new advancements and allied technologies must be made interoperable with existing systems, including required improvements in discrimination capabilities essential to the efficient employment of limited missile defense resources.

Intelligence, Surveillance, and Reconnaissance

Timely, useable situational awareness and intelligence analysis is essential to all military operations. Airborne, submarine, and space-based ISR capabilities all provide key indications and warning information to commanders facing an array of traditional adversaries, non-traditional threats, and challenging intelligence problems. For the past decade, ISR efforts focused primarily on meeting the expanding demand in the U.S. Central Command (USCENTCOM) area of responsibility. As overseas contingency operations change, DoD must carefully examine force requirements to ensure we organize, train, and equip a balanced force across the range of requirements, including anti-access environments and New START verification. An objective, multi-domain, capabilities-based architecture that improves the ability to identify requirements across geographic boundaries and the range of potential threats is essential to appropriately balancing risk against necessary programmatic, budgetary, and acquisition decision points.

USSTRATCOM's ISR efforts achieved significant resource efficiencies and shaped ISR capability decisions through initiatives like the ISR Force Sizing Construct project, the High Altitude Transition study, the Synoptic Operational Area Reconnaissance Study, and the Mobile Nuclear Air Sampling Study. USSTRATCOM also successfully advocated for a critical USCENTCOM ISR capability—designed and executed in approximately 30 months and at a lower cost than traditional acquisition processes. The Services and intelligence community must continue to strive for better integration in order to reach greater efficiencies—not only for the collection platforms themselves but also across the still-limited processing, exploitation, and dissemination architecture needed to transform collections into actionable intelligence.

Combating Weapons of Mass Destruction

Another mission area requiring sustained attention is CWMD, since the pursuit of WMD by violent extremists and their proliferation to additional states remains the primary threat to the United States, our allies, and our partners. USSTRATCOM received the responsibility to synchronize DoD CWMD activities in 2005 and has made discouraging, detecting, deterring, and, if necessary, defeating these threats a priority for theater operations and strategic deterrence. Some actors seek nuclear, biological, or chemical weapons to coerce their neighbors or to deter U.S. intervention in regional conflicts. Others may seek such weapons to use them in terrorist attacks or as weapons of war. Diffuse networks of non-state entities, secretive state sponsors, shell corporations, and terrorist-financed transactions challenge our intelligence organizations to develop comprehensive, accurate, and actionable assessments that enable global CWMD. USSTRATCOM continues to pursue further national CWMD capability improvements with interagency partners to coordinate CWMD objectives, plans, and activities.

Among current and future CWMD enhancements are technological improvements to detect, analyze, and assess WMD developments. The 2010 QDR affirmed the need to enhance National Technical Nuclear Forensics capabilities to increase nuclear threat attribution and to deter those considering nuclear weapons transfer or use. In the past year, the USSTRATCOM Center for CWMD (SCC WMD) embedded Proliferation Security Initiative activities within U.S. Africa Command, U.S. Central Command and U.S. Southern Command exercises and supported planning and funding efforts to expand exercise participation and training synchronization across geographic combatant commands. Finally, SCC WMD collaboratively operates the Interagency Combating Weapons of Mass Destruction Database of Responsibilities, Authorities, and Capabilities (INDRAC) System with the Defense Threat Reduction Agency. INDRAC provides a strategic level information reference resource to inform CWMD operations, planning, advocacy, training, and exercises across the government.

In the 2010 QDR, the Secretary of Defense directed DoD to establish a Joint Task Force Elimination Headquarters to "better plan, train, and execute WMD-elimination operations...with increased nuclear disablement, exploitation, intelligence, and coordination capabilities." Last December, Secretary Gates tasked USSTRATCOM to execute this task and stand up a Standing Joint Force Headquarters for Elimination of WMD with "standing exploitation and intelligence cells in order to plan, train for, and execute global WMD elimination operations." USSTRATCOM is currently analyzing the requirements necessary to implement the Secretary's direction.

CONCLUSION

Great challenges lie ahead of the United States and USSTRATCOM, but so too do great opportunities. The Command is dedicated to being an effective steward of taxpayer resources while maintaining a strategic force structure ready and able to deter aggression, preserve U.S. freedom of action, and defeat adversaries when necessary. The uncertainty inherent in today's compleç multi-domain security environment requires that we summon our best efforts to develop and deploy the plans, systems, and forces needed to sustain America's deterrent, ensure unfettered access to and through space and cyberspace, and win the dynamic joint fight. I look forward to working with Congress as we pursue these priorities together, and I appreciate your support and counsel in the months and years ahead. Thank you again for the opportunity to be here today.

In: Combatant Commands in the United States Armed Forces ISBN: 978-1-61942-924-6
Editors: Quincy N. Butler and Harris A. Simmons

Chapter 10

STATEMENT OF GENERAL DUNCAN J. MCNABB, USAF COMMANDER, UNITED STATES TRANSPORTATION COMMAND BEFORE THE SENATE ARMED SERVICES COMMITTEE ON THE STATE OF THE COMMAND*

April 7, 2011

INTRODUCING THE UNITED STATES TRANSPORTATION COMMAND MISSION/ORGANIZATION

The United States Transportation Command (USTRANSCOM) is a Total Force team of Active Duty, Guard, Reserve, civilian, commercial partners and contractors who lead a world class Joint Deployment and Distribution Enterprise (JDDE). Through our component commands, the Army's Military Surface Deployment and Distribution Command (SDDC), the Navy's Military Sealift Command (MSC), the Air Force's Air Mobility Command (AMC) and our commercial partners, we execute military and commercial transportation, distribution process integration, terminal management, aerial refueling and global patient movement across the full range of military and humanitarian operations. We operate effectively and efficiently to deploy and sustain the warfighter...and 2010 was a banner year. Together with our Service, Combatant Command, Interagency and Coalition partners, the USTRANSCOM team responded superbly to the President's direction to increase forces by 30,000 in Afghanistan, to drawdown forces to 50,000 in Iraq and to an unprecedented series of world events and natural disasters. Whether delivering combat power to Afghanistan through logistics or humanitarian relief to the people of Pakistan, Haiti and Japan, our team kept our promises and delivered on time, on target and at best value to the taxpayer.

* These remarks were delivered as testimony given on April 7, 2011. General Duncan J. Mcnabb, USAF Commander, United States Transportation Command before the Senate Armed Services Committee on the State of the Command.

Our People

When faced with immediate and long-term world events, the men and women of our superb TRANSCOM team overcome colossal obstacles to support our Nation's objectives with world-class logistics. In the history of the command, we've never had a better group of experienced, dutiful and enthusiastic individuals to fulfill the promises we make to the Combatant Commanders and warfighters. As challenges arise, this team ignites their talent, insight, flexibility, and ingenuity to swiftly design a way to deliver, whenever, wherever. Simply put, the spirit and work of the people who make up the Total Force TRANSCOM team has put the command on the world stage as the best of the best for delivering global logistics superiority.

Supporting Global Operations

Over the past year, USTRANSCOM components moved near-record quantities of cargo and supplies and tens of thousands of service personnel to all parts of the globe. AMC and our Civil Reserve Air Fleet (CRAF) commercial partners airlifted more than 2 million passengers and 848,000 tons of cargo, while AMC's aging air refueling fleet delivered 202 million gallons of fuel to U.S. and coalition aircraft. Equally impressive, MSC, SDDC and our commercial sealift partners moved over 3.8 million tons of cargo worldwide. Finally, MSC's tankers delivered 1.5 billion gallons of fuel to support operations around the world.

Support to USCENTCOM

USTRANSCOM continued its focus on supporting operations in the United States Central Command (USCENTCOM) Area of Responsibility (AOR). In 2010, we deployed and redeployed 48 Brigade Combat Teams, 75,000 Air Expeditionary Forces, 12 Security Force packages, and moved Marine Expeditionary, Stryker and Combat Aviation Brigades. The centerpiece of our efforts this year was the team's successful and on-time deployment of the 30,000 surge force into Afghanistan and drawdown from 130,000 to 50,000 service members in Iraq – both completed on our target date of 31 August 2010.

In Iraq, and in close coordination with USCENTCOM, we began the drawdown in earnest in May 2010 and redeployed 9,000 service members per week. We were able to achieve this remarkable volume of passenger movement by leveraging an additional strategic redeployment hub in Al Asad, Iraq. The addition of Al Asad to our existing hub in Kuwait, and the great flexibility and responsiveness of our CRAF partners, allowed us to nearly double our capacity to move military passengers and meet the President's 31 August 2010 deadline.

In Afghanistan, the lack of developed and robust infrastructure required close collaboration with other Combatant Commanders, the Joint Staff and our inter-agency partners to further expand capacity of our existing ground lines of supply in both Pakistan and the Northern Distribution Network, to increase through-put at airfields and to add further seaport and airfield capacity. The team was successful in our capacity-building efforts. In eight months, USTRANSCOM's components and our commercial partners delivered 30,000 troops and 60,000 tons of equipment and supplies to Afghanistan by 31 August 2010, again meeting the President's direction to increase the force by the end of August.

At the height of the drawdown in Iraq and surge in Afghanistan, an almost monthly occurrence of world events and natural disasters took place. Each of these events carried significant transportation and logistics implications, which challenged the USTRANSCOM team and our partners to both meet emerging requirements and stay on time in Iraq and Afghanistan. This demonstrated USTRANSCOM's flexibility to use its Total Force and U.S Flagged carriers to surge capacity to meet worldwide requirements.

In January of 2010, a 7.0 magnitude earthquake struck Haiti and required an immediate response by all three TRANSCOM components and our commercial sealift partners in support of USSOUTHCOM and USAID. The team reacted magnificently providing aid and relief supplies within hours and days of the devastation. In February, a coup in Kyrgyzstan disrupted operations at our primary Operation ENDURING FREEDOM personnel transit hub at Manas Transit Center in the Kyrgz capital of Bishkek. This required us to quickly reroute thousands of military passengers to Kuwait. Once again the system responded immediately, and deployment operations to Afghanistan continued without delay. Almost immediately after the return to normal operations in Kyrgyzstan, we were challenged in March by the month-long volcanic eruptions in Iceland which dramatically affected most, and sometimes all, of European airspace.

Again, TRANSCOM and our partners responded immediately and rerouted cargo, passenger and aeromedical evacuation missions around the affected airspace and vital cargo and passenger missions arrived with little or no delay. Finally, and as the drawdown in Iraq and the surge in Afghanistan were coming to a close, Pakistan experienced unprecedented devastation due to heavy rains which flooded over one fifth of the country and displaced 20 million people. USTRANSCOM, in coordination with USCENTCOM, responded by delivering 400,000 meals within 72 hours to those in greatest need. We also mobilized contingency response elements and delivered helicopters and boats to distribute over 8,500 tons of aid to remote provinces.

Improving throughput and expanding capacity in our surface networks which supply Afghanistan has again been a centerpiece of our efforts in 2010. The Northern Distribution Network (NDN) remains a priority for USTRANSCOM, and we have delivered over 30,000 containers via this network. In 2010, we added two additional routes through the Baltics and Central Asia and continue to improve the processes, facilitating a faster, less costly cargo flow.

In addition to the NDN improvements, we added capacity in intermodal Persian Gulf locations. Realizing we needed more capacity to support the surge of forces into Afghanistan and the movement of thousands of Mine Resistant Ambush Protected Vehicles (MRAP), the team worked closely with CENTCOM and the Department of State to secure access to additional airfields and seaports in the Persian Gulf. Using a concept called multi-modal operations, we moved large volumes of cargo and thousands of vehicles by sea to locations in closer proximity to the CENTCOM area of operations, by truck from the seaports to the nearby airfields and then by air to Afghanistan. This concept was used with great success throughout 2010 as we moved almost 7,000 MRAP and MRAP all-terrain vehicles to Afghanistan. Utilizing the combination of air, land and sea modes of transportation resulted in increased velocity, better utilization of aircraft and ultimately reduced costs by almost $400M in 2010.

The mountainous terrain and poor infrastructure in Afghanistan also required an increased reliance on aerial delivery, and 2010 was a record year with over 60 million pounds of cargo delivered by airdrop -- almost double the amount in 2009. The mobility air force continues to add new capabilities such as Low-cost, Low-altitude (LCLA) Delivery and Low-cost Aerial Delivery System (LCADS). USTRANSCOM is also exploring a high-speed container delivery system capability for the C-130J and C-17. This capability will improve both the survivability of the aircraft and accuracy of aerial delivery.

Finally, threats to our operations in USCENTCOM are not isolated to Afghanistan. Somali-based pirates continue to hazard our commercial sealift partners. USTRANSCOM and Military Sealift Command continue to actively engage with the Maritime Administration, the Coast Guard, the Navy and our industry partners to further reduce the vulnerability of the U.S. flag commercial fleet. One tactic which has been extremely successful is the mindful use of contracted armed security teams aboard U.S. flag commercial vessels.

SUPPORT TO OTHER COMBATANT COMMANDS

The United States Southern Command (USSOUTHCOM) was a very active AOR. As previously mentioned, Operation UNIFIED RESPONSE provided relief to Haiti after a massive earthquake on January 12, 2010, and USTRANSCOM was a key partner in that effort. Within two days of the earthquake, USTRANSCOM deployed its Joint Task Force – Port Opening (JTFPO) units to re-establish airfield and seaport operations after the earthquake had largely destroyed Haiti's existing infrastructure. This rapid response allowed USTRANSCOM and our commercial partners to deliver over 400,000 tons of lifesaving cargo, more than 2.5 million meals, and over 5 million liters of water to Haitians in need. Further, USTRANSCOM aeromedical evacuation teams safely moved 361 critically injured earthquake victims.

Support for the NATO security presence in Kosovo continued in the United States European Command (USEUCOM) AOR. USTRANSCOM moved over 2,500 service members into the Balkans in support of that mission, and provided strategic airlift support to five major USEUCOM and NATO exercises.

Our support to United States Africa Command (USAFRICOM) continued to increase this year. We moved 1,476 service members and more than 16,000 tons of cargo for Exercise AFRICAN LION, USAFRICOM's largest combined exercise.

This past year, we began a close partnership with the newly constituted United States Cyber Command to improve information operations security and to counter cyber threats to our networks. Winning the cyber fight is critical to safeguarding the systems and information which enable our global logistics network to operate.

USTRANSCOM not only moves cargo and people, we move information as well. Our vigilance will only increase as we work with our partners to defend our networks and information in this new battlespace.

In the United States Northern Command (USNORTHCOM) AOR, USTRANSCOM deployed three Modular Aerial Spray System-equipped C-130 aircraft and over 60 personnel to begin immediate oil dispersant operations over the Deepwater Horizon oil spill in the Gulf of Mexico. Throughout the 33-day deployment, the team flew 92 sorties and released more than 156,000 gallons of oil dispersant over the spill. Additionally, USTRANSCOM airlifted over 259 tons of booms, skimmer boats and other oil spill containment equipment to support the clean-up.

USTRANSCOM also supported a series of USNORTHCOM exercises which provided realistic homeland defense and Defense Support to Civil Authorities training for joint and interagency partners throughout the federal government.

The United States Pacific Command (USPACOM) AOR is the largest and most diverse in the world, and USTRANSCOM supported operations from Alaska to Antarctica and around the Pacific Rim and the Indian Ocean. In support of the National Science Foundation's Operation DEEP FREEZE (ODF), USTRANSCOM delivered more than 3,250 passengers, 10,000 tons of cargo, and 5.1 million gallons of fuel into McMurdo Station, Antarctica. In the Pacific Rim, USTRANSCOM provided humanitarian assistance and disaster response in the wake of Typhoon Fanapi in the Philippines and the Mount Merapi volcanic eruptions in Indonesia. Additionally, USTRANSCOM transported more than 687 passengers and 13,300 tons of cargo for Operation ENDURING FREEDOM—Philippines.

Support to the Warfighter

Global patient movement remains one of USTRANSCOM's most demanding missions as it requires 100 percent accuracy. Thanks, in large part, to rapid global patient movement, the wounded warrior survival rate has increased from 75 percent a decade ago to over 92 percent today. The survival rate increases to 98 percent if a wounded warrior makes it to a hospital alive. In 2010, USTRANSCOM completed more than 26,600 patient movements, all without incident. Additionally, USTRANSCOM rapidly deployed patient movement expertise all over the globe. Within 48 hours of the earthquake in Haiti, USTRANSCOM patient movement personnel were on-scene coordinating the movement of critically injured patients. When the Icelandic volcano erupted, USTRANSCOM rapidly altered aeromedical evacuation flight routing through the Mediterranean to ensure uninterrupted return of our wounded warriors from Southwest Asia.

USTRANSCOM support to the warfighter is not reserved solely for the battlefield. We recognize the need to care for families, including the effective and efficient movement of household goods. To that end, USTRANSCOM continued to field the Defense Personal Property System (DPS). DPS is a next generation, web-based system for management of personal property shipments and is helping to provide the best-value move for DOD families.

DPS executed more than 338,000 shipments in FY10 and will soon be used for nearly all shipments of household goods for DOD military and civilian personnel and their families.

Leading the Joint Deployment and Distribution Enterprise

The ability of the U.S. to project and sustain our forces over global distances is one of our nation's greatest asymmetric advantages. Our success depends on our ability to synchronize deployment distribution planning and execution across DOD, the regional Combatant Commands, the Services and our coalition and interagency partners. To that end, upon the President's approval, USTRANSCOM is poised to assume an additional Unified Command Plan (UCP) mission as the "Global Distribution Synchronizer." In collaboration with our partners, this new mission will enable us to shape the distribution environment and gain greater access to distribution lanes that cross multiple theaters to project and sustain forces globally. Collectively, we will "knit the seams" between multiple Combatant Command theater distribution campaign plans and create a more robust and adaptive distribution network that reduces operational risk.

ENHANCING DOD SUPPLY CHAIN MANAGEMENT

USTRANSCOM is leading the transformation of the DOD supply chain through a series of Distribution Process Owner Strategic Opportunities (DSO) initiatives. These include five major opportunities to enhance readiness, improve velocity and reduce DOD supply chain cost.

Strategic Surface and Air Optimization lower the cost of shipments by consolidating surface cargo into single containers, as well as modifying pallet build rules and using —less-thanplaneload|| commercial freight services for air cargo. Through Strategic Network Optimization and Supply Alignment, USTRANSCOM optimizes the number, location and function of supply chain nodes to increase distribution effectiveness by positioning selected materiel in forward locations to reduce reliance on high-cost air transportation. Finally, through process improvement, USTRANSCOM increases velocity across the enterprise by identifying and removing —dead time|| throughout the supply chain process.

These initiatives are generating benefits to the warfighter by delivering higher levels of service at lower costs. To date, the DSO initiatives have generated savings of $80 million and an improvement in delivery time of up to 34 percent.

Global distribution efficiency begins with the best value movement of DOD freight in the Continental United States (CONUS). This is the purpose of the Defense Transportation Coordination Initiative (DTCI). Using commercial best practices, DTCI improves the reliability, predictability, and efficiency of DOD materiel moving in the CONUS. Thus far, DTCI has saved $182 million and meets or exceeds goals for on-time pickup, reduced damage claims and small business participation.

Business process reengineering and Corporate Services Vision are at the heart of USTRANSCOM's transformation efforts. Agile Transportation for the 21st Century (AT21) is one such effort which uses industry best practices and government and commercial off-the-shelf optimization and scheduling tools to deliver best value, end-to-end distribution and deployment solutions. Business process reengineering improves transportation planning, forecast accuracy and on-time delivery of forces and sustainment to Combatant Commanders at a lower cost to the Services. Equally important, our Corporate Services Vision aligns information technology systems with these reengineered business processes with a one-stop information technology shop. We expect our AT21 to deliver a significant return on investment. We are also investigating industry-leading collaborative technologies, computer gaming, and social networking innovations to provide additional capability.

In cooperation with the Defense Logistics Agency (DLA), USTRANSCOM is improving visibility across the supply chain through the Integrated Data Environment/Global Transportation Network Convergence (IGC) initiative.

The purpose of IGC is to collect supply, transportation and deployment data from disparate systems and allow access to that data from anywhere in the world. This capability provides warfighters access to real time, actionable logistics information and allows them to make informed decisions.

As DOD's lead proponent for radio frequency identification (RFID) and related automatic identification technology (AIT), our focus is on implementing the proper technologies to enhance supply chain business processes. While active RFID remains the primary AIT enabling in-transit visibility, this year we used satellite technology in high-threat areas where it is necessary to have near real-time location of critical assets while in transit. Additionally, we are exploring the use of sensor technology to enhance security of high-value cargo.

Container intrusion sensors also provide a force protection layer, alerting the warfighter to take extra safety precautions due to a container breach. Finally, passive RFID will enable supply chain process improvements such as increased inventory accuracy and decreased time to receive, store and issue material.

Realigning the Organization and Personnel

Another notable achievement in 2010 was the completion of the Base Realignment and Closure (BRAC) relocation of SDDC to Scott Air Force Base. The cornerstone of the project was the $130 million BRAC-funded MILCON facility, which was operational well ahead of schedule. Coincident to the co-location of SDDC with AMC and USTRANSCOM, we reengineered business processes and consolidated functions to achieve operational and fiscal efficiencies.

Operational benefits include fused operations and intelligence centers, a joint billing center and consolidated acquisition and analysis centers of excellence. The results have been impressive -- 470 fewer billets, a 20 percent reduction in contracts, elimination of two leased buildings, and a projected cost saving of $1.2 billion over 20 years.

USTRANSCOM continues to emphasize professional development of our human capitol. Our focus is to develop joint logisticians who can perform core functions inherent in the command's Distribution Process Owner (DPO) responsibilities. Innovative logistics solutions, like the NDN, require new ways of thinking about supply chains, developing sustainable infrastructure, running distribution networks in remote geographies, and building enduring international relationships. To develop that intellectual capital and critical thinking ability in our people, we added industry-leading courses from the University of Wisconsin, Pennsylvania State University, Carnegie Mellon University, Stanford University, University of Tennessee and Massachusetts Institute of Technology to our professional development curriculum.

Maintaining Air Mobility Readiness

Rapid global mobility is a key enabler to the effectiveness of the joint force. The ability to mobilize forces and materiel within hours, rather than days or weeks, depends on versatile, ready and effective air mobility forces.

In order to maintain our decisive global mobility advantage, we must have a viable tanker fleet. Therefore, the re-capitalization of the tanker fleet remains my top acquisition priority. The KC-46A will fulfill its primary refueling role and also have the flexibility to contribute to an array of mobility missions. It will dramatically improve our ability to do the air refueling mission and allow us to make the whole air mobility system much more efficient.

Likewise, our national defense strategy requires a viable fleet of about 300 strategic airlift aircraft. The C-17 Globemaster III will continue to be our premier airlifter, and our modernized C-5s are achieving their expected levels of mission readiness. However, in order to achieve the correct mix of C-17 and C-5 aircraft, and take full advantage of our critical aircrews and maintainers, the Air Force should be given the authority to retire the oldest, least capable C-5s.

C-17s will continue to meet USTRANSCOM's future requirements through currently funded purchases, upgrade programs, and fleet rotation. New C-17s arrive with improvements which increase the reliability of the weapon system. Older aircraft enter into the Global Reach Improvement Program to increase sustainability and reliability. Aircraft located in corrosive and training environments are monitored and analyzed for stress and rotated to maintain structural integrity of the fleet. Furthermore, two additional reserve component units will take on the C-17 mission as they retire their C-5s.

The C-5 is critical to our oversized and outsized air cargo capability. C-5 fleet management has two main focus areas: C-5 reliability and C-5A retirements. The Reliability Enhancement and Re-Engining Program (RERP) will increase the mission capable rate (MCR) of the C-5 fleet. All C-5B and C models and one C-5A model aircraft will undergo RERP resulting in a total of 52 C-5Ms in the inventory. Additionally, the new maintenance processes changed our focus from —fly to fail|| on major components to preventative replacement. This has reduced the number of C-5s stranded off-station awaiting parts and will result in a seven percent increase in MCR. Finally, C-5A retirements will improve aircraft availability by removing maintenance intensive jets from the fleet and will allow us to focus our maintenance personnel and resources on the right sized fleet.

The intra-theater workhorse supporting the warfighter is the C-130. The Mobility Capabilities and Requirements Study-2016 (MCRS-16) determined that 335 C-130s are required to perform general support intra-theater airlift missions. Follow-on analysis of the direct support mission determined that 20 additional C-130s and the 38 C-27Js already in the program can perform the direct support mission. Air Mobility Command's assessment is that a total of 355 C-130s and 38 C-27Js, in both general and direct support roles, will support the warfighter.

Finally, I support the Air Force's plan to acquire up to 134 C-130Js, modernizing 221 legacy C-130s with the Avionics Modernization Program (AMP), and fielding 38 C-27Js.

Our mobility aircraft routinely operate in threat areas across the spectrum of conflict from humanitarian relief to combat resupply. To operate safely in these environments, I strongly support continued defensive systems upgrades. These upgrades include equipping aircraft with the Large Aircraft Infrared Countermeasures system and beginning development of the Advanced Situational Awareness and Countermeasures capability for operations in low-medium threat environments.

Operational Support Airlift (OSA) is another key component of our mobility force. Our senior leaders require immediate airlift to carry out military and other missions in a fluid strategic environment. It is important that we not only right-size and modernize the OSA fleet, but we need to develop a management system with a common multi-Service database and operational picture. The goal is to achieve real-time visibility of worldwide senior leader and OSA movements to enable USTRANSCOM and Geographic Combatant Commanders to exercise command and control of the OSA fleet within their area of responsibility.

Just as command and control of OSA assets is critical, the leaders aboard the aircraft must be able to communicate while they travel. This requires secure, reliable communications for U.S. national leaders anywhere on the globe. USTRANSCOM continues to work with other DOD and U.S. Government stakeholders through the National Leadership Command Capability Executive Management Board to provide the proper level of Senior Leadership Command, Control and Communications – Airborne (SLC3S-A). The Senior Leader In-transit Conference Capsule (SLICC) is a transportable "office in the sky" that provides a secure, interconnected meeting place for Senior Leaders traveling aboard C-17, KC-10 and C-130 aircraft. Designed with flexibility and affordability in mind, the SLICC promises to reduce operating costs by enabling Combatant Commanders to support DV missions using available theater assets and regularly scheduled airlift routes.

The Civil Reserve Air Fleet (CRAF) is a critical component in our ability to rapidly deploy forces and equipment. In this past year, through discussions with our commercial industry partners, we made the most sweeping changes to the CRAF program in 15 years. These changes will result in a stronger, more viable program and address Congressional mandates to improve predictability of DOD commercial requirements and incentivize carriers to use modern aircraft. Perhaps the most profound change is the implementation of a —flyer bonus|| which, for the first time in the history of the CRAF, rewards peacetime mission participation in addition to the traditional reward for wartime commitment. Our plan for FY12 is to amend the flyer bonus to provide increased reward to those carriers who fly peacetime CRAF missions with modernized aircraft.

To ensure the CRAF partnership remains strong, USTRANSCOM created the Executive Working Group (EWG), modeled after a similar venue with our sealift partners. The CRAF EWG brings together USTRANSCOM and AMC leaders with Chief Executive Officers, Presidents, and other representatives of the commercial airline industry to discuss issues with the CRAF program. Since its first meeting in March 2010, this group has met six times and proved instrumental in crafting the changes to the FY11 contract. The CRAF EWG will continue to meet on a regular basis to discuss additional changes to this vital program.

Maintaining Sealift Readiness

Sealift is the primary means for delivering ground forces and sustainment during major combat operations, and has been responsible for delivering over 90 percent of all cargo to Afghanistan and Iraq. Because of the superb volunteer participation of commercial U.S.-

Flagged vessels in the Maritime Security Program (MSP), we did not have to activate a single ship in the Surge Fleet or the Ready Reserve Force (RRF) to meet the President's aggressive timeline for the surge and drawdown of forces in Afghanistan and Iraq—a remarkable achievement.

The large, medium speed, roll-on, roll-off ships (LMSRs) in the Surge Fleet, the vessels of the RRF and the commercial U.S. Flag Fleet in the MSP and Voluntary Intermodal Support Agreement (VISA) are all required to meet the Nation's strategic sealift requirements. While cargo preference laws and national defense sealift policies ensure the viability of the U.S. flag commercial fleet, we must also continue to keep the Surge Fleet and Ready Reserve Force vessels at an equal state of readiness as well as our citizen mariners who man these vessels during operations in USCENTCOM and around the world. In fact, the Maritime Administration is conducting a thorough study of cargo preference laws to ensure that they most effectively support the delicate balance of commercial viability and readiness which is so critical to our sealift capability.

USTRANSCOM's partnership with the U.S. commercial sealift industry and the Department of Transportation has been vitally important in developing new routes for conveying cargo around the globe – particularly to regions with undeveloped infrastructure. Through programs like the Maritime Security Program (MSP), the Voluntary Intermodal Sealift Agreement and the Voluntary Tanker Agreement, the Department of Defense gains access to U.S. commercial capabilities and transportation networks while ensuring the continued viability of both the U.S. flag fleet and the pool of citizen mariners who man those vessels. Last year, Congress ensured the continuation of the MSP by extending it an additional 10 years to 2025. We look forward to working with Congress and this committee to refine this program between now and the MSP implementation date in 2015.

We also work closely with the DLA Energy office to meet DOD's fuel requirements. On October 7, 2010, MSC took operational control of the first of two U.S.-built, U.S.-flagged State-Class tanker vessels. These new double-hulled 331,000-barrel ships will provide vital sealift capabilities. The MV EMPIRE STATE and her sister ship, MV EVERGREEN STATE, will carry refined petroleum products between commercial refineries and DOD storage and distribution facilities worldwide. While these two new tankers increase the total number of U.S.- flagged tankers with international trade rights to five, our nation would benefit greatly from even further increases in U.S.-flagged tanker capacity. Additionally, the Maritime Administration is currently examining the feasibility and potential benefit of developing a strategic access program outside of MSP solely for tankers.

Finally, we look forward to working with the Navy and Combatant Commanders to fully realize the logistics capabilities of the Joint High Speed Vessel across all its missions from Theater Security Cooperation to its use as an operational and tactical logistics platform.

Maintaining Surface Readiness

USTRANSCOM depends on our en route structure to rapidly support theater COCOMS. On August 9, 2010 USTRANSCOM submitted its inaugural En Route Infrastructure Master Plan (ERIMP) 2010 to the Chairman of the Joint Chiefs of Staff. The purpose of the ERIMP is to guide the en route infrastructure investment decisions necessary to ensure we support the regional Combatant Commander Theater Campaign and Theater Posture Plans. The ERIMP frames the en route strategy by identifying our most important enterprise-wide infrastructure requirements. It includes recommendations from the Combatant Commands and takes an enterprise approach to global deployment and distribution.

Because en route infrastructure is key to global logistics, Rota, Spain; Camp Lemonier, Djibouti; Souda Bay on the island of Crete; and Guam remain USTRANSCOM near-term priorities as strategic locations. Similar to our multi-modal MRAP movement through the Persian Gulf, movement of Combat Aviation Brigade helicopters by ship to the intermodal port at Rota, Spain, then flying them into Afghanistan significantly increases velocity. Camp Lemonier is the only en route node on the African continent with enduring presence; an unrefueled C-17 can reach two-thirds of the continent from Djibouti. Funding the expansion of the aircraft parking area and the taxiway will help ensure Camp Lemonier's continued viability as a critical strategic intermodal location. Located on the island of Crete in the central Mediterranean Sea, Souda Bay is key to the JDDE en route mission due to its proximity to the Black Sea, the Middle East, and Africa. Funding for a multi-access road, aircraft parking expansion, an air operations complex, and the Marathi logistics facility will help tremendously as we support three separate Geographic Combatant Commanders from Souda Bay. Guam is our intermodal crown jewel in the Pacific. The USTRANSCOM-led 2009 Global En Route Infrastructure Steering Committee identified two necessary military construction (MILCON) projects at Anderson Air Force Base, Guam: an Air Freight Terminal Complex and Air Passenger Terminal/Joint Personnel Deployment Center. Each of these projects will greatly enhance the effectiveness, efficiency and safety of the passenger and the cargo handling process.

At home, our highways, railroads, and ports for national defense programs work in partnership with the Department of Transportation (DOT) to identify DOD's requirements for the civil sector infrastructure between our installations and ports and integrate these requirements into the civil sector planning cycle. These programs play a key role in ensuring our infrastructure in CONUS is ready to support DOD's deployment and distribution needs.

DEVELOPING NEW CAPABILITIES

Global logistics is an incredibly fluid business. As the geopolitical, diplomatic and operational landscape changes, USTRANSCOM understands the need for innovative mobility and distribution strategies, processes and technology.

Our Research, Development, Test, and Evaluation Program (RDT&E) searches for these emerging enablers to support the future force. Through this program, we leverage new technologies, cutting edge business processes and innovative logistics strategies to address mobility challenges before they arise.

Given the poor infrastructure in Afghanistan, much of our work focuses on new methods of aerial delivery. We are working with the Natick Soldier Center to develop the Helicopter Sling Load of the Joint Precision Airdrop System. The project integrates elements from various airdrop programs into a new capability that will allow for the delivery of payloads from a helicopter cargo hook. The system also ensures rapid distribution for both delivery over land and ship-to-shore applications.

We also continue to invest in intelligent unmanned aircraft technology to autonomously deliver critical supplies to forward points of need. Our intent is to address extended lines of communication susceptible to weather, degraded road conditions and enemy threats, such as improvised explosive devices, ambush, and sabotage.

Through the Joint Capabilities Technology Demonstration program, we are supporting High Speed Container Delivery Systems (HSCDS) to improve airdrop accuracy, increase tonnage dropped, and enhance survivability of airlift and aircrews. HSCDS is a high-speed, low altitude airdrop system that provides the warfighter more cargo, more often and with more accuracy than any comparable delivery system. It optimizes aircraft threat avoidance and tactical maneuverability while enhancing our ability to deliver vital cargo to small combat units at the point of need.

Hybrid airships can revolutionize logistics by moving the supply chain above the battle space to deliver large volumes of cargo directly to the point of effect, without the need for an airfield or roads. By delivering directly, hybrid airships bypass many supply-chain "touches," thereby reducing cost and risk. The continued exploration of the hybrid airship concept is essential to support future operations in austere and infrastructure-challenged locales where USTRANSCOM will likely be required to support military or humanitarian relief operations.

The Joint Recovery and Distribution System (JRADS) achieved tremendous success in its first demonstrations with the 101st Sustainment Brigade. The JRADS technology enables quick and efficient recovery of damaged MRAPs on the battlefield, which minimizes troop exposure to attack when recovering damaged vehicles.

USTRANSCOM is committed to innovative sealift solutions as well. Sea basing is one such innovation which affords alternatives to the traditional use of seaports of debarkation. It enables discharge, reception, staging and assembly at sea; and interfaces with both organic and commercial sealift assets. The Large Vessel Interface Lift-On/Lift-Off system provides the ability to load and offload containers between ships at sea with precision and in much higher sea states than is currently possible. In April 2010, the technology was successfully demonstrated in the Gulf of Mexico between Ready Reserve Fleet vessels SS FLICKERTAIL STATE and MV CAPE TEXAS. Together with the Navy, we will further develop this technology.

The Joint High Speed Vessel represents a transformational sealift capability. Bridging the traditional gap between high-speed, low-capacity airlift and low-speed, high-capacity sealift, it affords the promise of enhanced logistic response to military and civil contingencies around the globe. Forward deployment of the vessel in combination with warehoused stocks of equipment and supplies will leverage its speed and capacity to quickly deliver needed cargo.

FISCAL STEWARDSHIP

Good stewardship of the taxpayers' dollars is a USTRANSCOM hallmark. In everything we do, we are always effective, but mindful of cost. Since 2003, we and our enterprise partners have avoided costs in excess of $5.2 billion through transformational distribution initiatives, improving inventory and transportation alignment, optimizing strategic air and surface processes and effectively utilizing multi-modal transportation solutions; all while improving end-to-end velocity and effectiveness.

Given our global mission, the AMC Fuel Efficiency Office has been laser-focused on increasing fuel efficiency. The result is a significant cost decrease to the customer and taxpayer. For example, information technology (IT) improvements such as Mission Index Flying for the C17 and C-5 reduce each sortie's fuel burn during mission execution. Other improvements, such as the Advanced Computer Flight Planning overlay, reduce the excess fuel carried which increases the amount of cargo the aircraft can carry -- again lowering the cost to the taxpayer.

USTRANSCOM continues to lead the certification effort for alternative fuels. Most aircraft in AMC's fleet are approved to fly on a synthetic blend of coal or natural gas-based Fischer-Tropsch fuel and JP-8 military grade jet fuel. The workhorse of AMC, the C-17, underwent flight tests and certification on a cutting-edge, renewable jet fuel blended with JP-8 this past August. In addition to allowing the DOD to target renewable sources, alternative fuels are more environmentally friendly than traditional jet fuel.

Final Thoughts

USTRANSCOM has the critical national responsibility to support the men and women who fight to preserve our liberty and security and to support those who provide lifesaving relief to those in need...and to do so in an global operating environment increasingly characterized by uncertainty, complexity, and rapid change. Looking to the future, USTRANSCOM, along with our enterprise partners, will continue to transform the Joint Deployment and Distribution Enterprise to meet this challenging new environment and continue to provide effective and best value support to our nation. We will always, always deliver.

INDEX

A

D

E

F

G

H

I

J

K

L

M

N

O

P

Q

R

S

T

U

V

W

Y

Z